BEHIND H

BEHIND HUMAN ERROR

Second Edition

David D. Woods
Sidney Dekker
Richard Cook
Leila Johannesen
&
Nadine Sarter

ASHGATE

© David D. Woods, Sidney Dekker, Richard Cook, Leila Johannesen and Nadine Sarter 2010

All rights reserved. No part of this publication may be reproduced, stored in a retrieval system or transmitted in any form or by any means, electronic, mechanical, photocopying, recording or otherwise without the prior permission of the publisher.

David D. Woods, Sidney Dekker, Richard Cook, Leila Johannesen and Nadine Sarter have asserted their right under the Copyright, Designs and Patents Act, 1988, to be identified as the authors of this work.

Published by
Ashgate Publishing Limited
Wey Court East
Union Road
Farnham
Surrey, GU9 7PT
England

Ashgate Publishing Company
Suite 420
101 Cherry Street
Burlington
VT 05401-4405
USA

www.ashgate.com

British Library Cataloguing in Publication Data
Behind human error.
1. Fallibility. 2. Human engineering.
I. Woods, David D., 1952-
620.8'2-dc22

ISBN: 978-0-7546-7833-5 (hbk)
       978-0-7546-7834-2 (pbk)
       978-0-7546-9650-6 (ebk)

Library of Congress Cataloging-in-Publication Data
Behind human error / by David D. Woods ... [et al.]. -- 2nd ed.
    p. cm.
Includes bibliographical references and index.
ISBN 978-0-7546-7833-5 (hbk.) -- ISBN 978-0-7546-9650-6 (ebook)
1. Industrial accidents--Prevention. 2. Human-machine systems. 3. Human-computer interaction. 4. Cognition.
5. Errors. I. Woods, David D.,
1952-
HD7262.B387 2010
363.11'6--dc22

2010008469

Reprinted 2011

Printed and bound in Great Britain by the
MPG Books Group, UK

CONTENTS

List of Figures		*vii*
List of Tables		*ix*
Acknowledgments		*xi*
Reviews for Behind Human Error, Second Edition		*xii*
About the Authors		*xiii*
Preface		*xv*
PART I	**AN INTRODUCTION TO THE SECOND STORY**	
1	The Problem with "Human Error"	3
2	Basic Premises	19
PART II	**COMPLEX SYSTEMS FAILURE**	
3	Linear and Latent Failure Models	41
4	Complexity, Control and Sociological Models	61
5	Resilience Engineering	83
PART III	**OPERATING AT THE SHARP END**	
6	Bringing Knowledge to Bear in Context	101
7	Mindset	113
8	Goal Conflicts	123

PART IV	HOW DESIGN CAN INDUCE ERROR	
9	Clumsy Use of Technology	143
10	How Computer-based Artifacts Shape Cognition and Collaboration	155
11	Mode Error in Supervisory Control	171
12	How Practitioners Adapt to Clumsy Technology	191
PART V	REACTIONS TO FAILURE	
13	Hindsight Bias	199
14	Error as Information	215
15	Balancing Accountability and Learning	225
16	Summing Up: How to Go Behind the Label "Human Error"	235
References		*251*
Index		*269*

LIST OF FIGURES

Figure 1.1	The sharp and blunt ends of a complex system	9
Figure 1.2	At the sharp of a complex system	9
Figure 1.3	The sharp end of a complex system is characterized by how practitioners adapt to cope with complexity.	10
Figure 1.4	Hindsight bias simplifies the situation, difficulties, and hazards faced before outcome is known	14
Figure 2.1	The relationship between error recovery and outcome failure	27
Figure 3.1	Complex systems failure according to the latent failure model	51
Figure 4.1	The difference between the crew's chart on the morning of the accident, the actual situation (center) and the eventual result of the reconstruction (NFDC or National Flight Data Center chart to the right)	73
Figure 4.2	The structure responsible for safety-control during airport construction at Lexington, and how control deteriorated	73
Figure 4.3	A space of possible organizational action is bounded by three constraints: safety, workload and economics	74
Figure 8.1	Conflicting goals in anesthesia	127
Figure 10.1	This "Impact Flow Diagram" illustrates the relationship between the design-shaping properties of the computer as a medium, the cognition-shaping properties of representations in the computer medium, and the behavior-shaping properties of cognitive systems	157
Figure 10.2	Eight minutes before the explosion	163
Figure 10.3	The moment of the explosion	164
Figure 10.4	Four seconds after the explosion	165
Figure 10.5	Four minutes after the explosion	165
Figure 10.6	Depicting relationships in a frame of reference	166
Figure 10.7	Putting data into context	167
Figure 10.8	Highlighting events and contrasts	168
Figure 11.1	Example of multiple modes and the potential for mode error on the flight deck of an advanced technology aircraft	183
Figure 12.1	How practitioners cope with complexity in computerized devices	193

LIST OF TABLES

Table 1.1   The contrast between first and second stories of failure   7
Table 3.1   Correlations between the number of nonfatal accidents or incidents
            per 100,000 major US jet air carrier departures and their passenger
            mortality risk (January 1, 1990 to March 31, 1996)   44

ACKNOWLEDGMENTS

This book first came about as a monograph in 1994 commissioned by the Crew Systems Ergonomics Information and Analysis Center of the Human Effectiveness Directorate of the Air Force Research Laboratory. As the original monograph became increasingly difficult to obtain, we received more and more calls from the safety and human systems communities for an updated and readily accessible version. We would like to thank Guy Loft of Ashgate for his encouragement to develop the second edition, and Lore Raes for her unrelenting efforts to help get it into shape.

The ideas in this book have developed from a complex web of interdisciplinary interactions. We are indebted to the pioneers of the New Look behind the label human error – John Senders, Jens Rasmussen, Jim Reason, Neville Moray, and Don Norman for their efforts in the early 1980s to discover a different path. There have been many participants in the various discussions trying to make sense of human error over the last 30 years who have influenced the ideas developed in this volume: Charles Billings, Véronique De Keyser, Baruch Fischhoff, Zvi Lanir, Todd LaPorte, Gene Rochlin, Emilie Roth, Marty Hatlie, John Wreathall, and many others. A special thanks is due Erik Hollnagel. With his vision, he always has been at the ready to rescue us when we become trapped in the bog of human error.

Ultimately, we extend a special thanks to all who work at the sharp end for their efforts and expertise which keeps such complex and hazardous systems working so well so much of the time. We also thank the many sharp-end practitioners who donated their time and expertise to participate in our studies to understand how systems succeed but sometimes fail.

Reviews for
Behind Human Error, Second Edition

'This book, by some of the leading error researchers, is essential reading for everyone concerned with the nature of human error. For scholars, Woods et al provide a critical perspective on the meaning of error. For organizations, they provide a roadmap for reducing vulnerability to error. For workers, they explain the daily tradeoffs and pressures that must be juggled. For technology developers, the book offers important warnings and guidance. Masterfully written, carefully reasoned, and compellingly presented.'

Gary Klein, Chairman and Chief Scientist of Klein Associates, USA

'This book is a long-awaited update of a hard-to-get work originally published in 1994. Written by some of the world's leading practitioners, it elegantly summarises the main work in this field over the last 30 years, and clearly and patiently illustrates the practical advantages of going "behind human error". Understanding human error as an effect of often deep, systemic vulnerabilities rather than as a cause of failure, is an important but necessary step forward from the oversimplified views that continue to hinder real progress in safety management.'

Erik Hollnagel, MINES ParisTech, France

'If you welcome the chance to re-evaluate some of your most cherished beliefs, if you enjoy having to view long-established ideas from an unfamiliar perspective, then you will be provoked, stimulated and informed by this book. Many of the ideas expressed here have been aired before in relative isolation, but linking them together in this multi-authored book gives them added power and coherence.'

James Reason, Professor Emeritus, University of Manchester, UK

'This updated and substantially expanded book draws together modern scientific understanding of mishaps too often simplistically viewed as caused by "human error". It helps us understand the actions of human operators at the "sharp end" and puts those actions appropriately in the overall system context of task, social, organizational, and equipment factors. Remarkably well written and free of technical jargon, this volume is a comprehensive treatment of value to anyone concerned with the safe, effective operation of human systems.'

Robert K. Dismukes, Chief Scientist for Aerospace Human Factors, NASA Ames Research Center, USA

'With the advent of unmanned systems in the military and expansion of robots beyond manufacturing into the home, healthcare, and public safety, Behind Human Error is a must-read for designers, program managers, and regulatory agencies. Roboticists no longer have an excuse that the human "part" isn't their job or is too esoteric to be practical; the fifteen premises and numerous case studies make it clear how to prevent technological disasters.'

Robin R. Murphy, Texas A&M University, USA

ABOUT THE AUTHORS

David D. Woods, Ph.D. is Professor at Ohio State University in the Institute for Ergonomics and Past-President of the Human Factors and Ergonomics Society. He was on the board of the National Patient Safety Foundation and served as Associate Director of the Veterans Health Administration's Midwest Center for Inquiry on Patient Safety. He received a Laurels Award from *Aviation Week* and *Space Technology* (1995). Together with Erik Hollnagel, he published two books on Joint Cognitive Systems (2006).

Sidney Dekker, Ph.D. is Professor of human factors and system safety at Lund University, Sweden, and active as airline pilot on the Boeing 737NG. He has lived and worked in seven countries, and has held visiting positions on healthcare safety at medical faculties in Canada and Australia. His other books include *Ten Questions About Human Error: A New View of Human Factors and System Safety* (2005), *The Field Guide to Understanding Human Error* (2006), and *Just Culture: Balancing Safety and Accountability* (2007).

Richard Cook, M.D. is an active physician, Associate Professor in the Department of Anesthesia and Critical Care, and also Director of the Cognitive Technologies Laboratory at the University of Chicago. Dr. Cook was a member of the Board of the National Patient Safety Foundation from its inception until 2007. He counts as a leading expert on medical accidents, complex system failures, and human performance at the sharp end of these systems. Among many other publications, he co-authored *A Tale of Two Stories: Contrasting Views of Patient Safety*.

Leila Johannesen, Ph.D. works as a human factors engineer on the user technology team at the IBM Silicon Valley lab in San Jose, CA. She is a member of the Silicon Valley lab accessibility team focusing on usability sessions with disabled participants and accessibility education for data management product teams. She is author of "The Interactions of Alicyn in Cyberland" (1994).

Nadine Sarter, Ph.D. is Associate Professor in the Department of Industrial and Operations Engineering and the Center for Ergonomics at the University of Michigan. With her pathbreaking research on mode error and automation complexities in modern airliners, she served as technical advisor to the Federal Aviation Administration's Human Factors Team in the 1990's to provide recommendations for the design, operation, and training for advanced "glass cockpit" aircraft and shared the Aerospace Laurels Award with David Woods.

PREFACE

A LABEL

Human error is a very elusive concept. Over the last three decades, we have been involved in discussions about error with many specialists who take widely different perspectives – operators, regulators, system developers, probability reliability assessment (PRA) specialists, experimental psychologists, accident investigators, and researchers who directly study "errors." We are continually impressed by the extraordinary diversity of notions and interpretations that have been associated with the label "human error." Fifteen years after the appearance of the first edition of *Behind Human Error* (with the subtitle Cognitive Systems, Computers and Hindsight) published by the Crew Systems Information and Analysis Center (CSERIAC), we still see organizations thinking safety will be enhanced if only they could track down and eliminate errors.

In the end, though, and as we pointed out in 1994, "human error" is just a label. It is an attribution, something that people say about the presumed cause of something after-the-fact. It is not a well-defined category of human performance that we can count, tabulate or eliminate. Attributing error to the actions of some person, team, or organization is fundamentally a social and psychological process, not an objective, technical one. This book goes *behind* the label "human error" to explore research findings on cognitive systems, design issues, organizational goal conflicts and much more.

Behind the label we discover a whole host of complex and compelling processes that go into the production of performance – both successful and erroneous, and our reactions to them. Research on error and organizational safety has kept pace with the evolution of research methods, disciplines and languages to help us dig ever deeper into the processes masked by the label. From investigating error-producing psychological mechanisms in the early 1980s, when researchers saw different categories of error as essential and independently existing, we now study complex processes of cross-adaptation and resilience, borrow from control theory and complexity theory, and have become acutely sensitive to the socially constructed nature of the label "human error" or any language used to ascribe credit or blame for performances deemed successful or unsuccessful.

Indeed, the book examines what goes into the production of the label "human error" by those who use it, that is, the social and psychological processes of attribution and hindsight that come *before* people settle on the label.

The realization that human error is a label, an attribution that can block learning and system improvements, is as old as human factors itself. During the Second World War, psychologists were mostly involved in personnel selection and training. Matching the person to the task was considered the best possible route to operational success. But increasingly, psychologists got pulled in to help deal with the subtle problems confronting operators of equipment *after* they had been selected and trained. It became apparent, for example, that fewer aircraft were lost to enemy action than in accidents, and the term "pilot error" started appearing more and more in training and combat accident reports. "Human error" became a catch-all for crew actions that got systems into trouble. Matching person to task no longer seemed enough. Operators made mistakes despite their selection and training.

Yet not everybody was satisfied with the label "human error." Was it sufficient as explanation? Or was it something that demanded an explanation – the starting point to investigate the circumstances that triggered such human actions and made them really quite understandable? Stanley Roscoe, one of the eminent early engineering psychologists, recalls:

> It happened this way. In 1943, Lt. Alphonse Chapanis was called on to figure out why pilots and copilots of P-47s, B-17s, and B-25s frequently retracted the wheels instead of the flaps after landing. Chapanis, who was the only psychologist at Wright Field until the end of the war, was not involved in the ongoing studies of human factors in equipment design. Still, he immediately noticed that the side-by-side wheel and flap controls – in most cases identical toggle switches or nearly identical levers – could easily be confused. He also noted that the corresponding controls on the C-47 were not adjacent and their methods of actuation were quite different; hence C-47 copilots never pulled up the wheels after landing. (1997, pp. 2–3)

"Human error" was not an explanation in terms of a psychological category of human deficiencies. It marked a beginning of the search for systemic explanations. The label really was placeholder that said, "I don't really know what went wrong here, we need to look deeper." A placeholder that encouraged further probing and investigation. Chapanis went behind the label to discover human actions that made perfect sense given the engineered and operational setting in which they were planned and executed. He was even able to cross-compare and show that a different configuration of controls (in his case the venerable C-47 aircraft) never triggered such "human errors." This work set in motion more "human error" research in human factors (Fitts and Jones, 1947; Singleton, 1973), as well as in laboratory studies of decision biases (Tversky and Kahneman, 1974), and in risk analysis (Dougherty and Fragola, 1990).

The Three Mile Island nuclear power plant accident in the US in the spring of 1979 greatly heightened the visibility of the label "human error." This highly publicized accident, and others that came after, drew the attention of the engineering, psychology, social science, regulatory communities and the public to issues surrounding human error.

The result was an intense cross-disciplinary and international consideration of the topic of the human contribution to risk. One can mark the emergence of this cross-disciplinary and international consideration of error with the "clambake" conference on human error organized by John Senders at Columbia Falls, Maine, in 1980 and with the publication of Don Norman's and Jim Reason's work on slips and lapses (Norman, 1981; Reason and Mycielska, 1982).

The discussions have continued in a wide variety of forums, including the Bellagio workshop on human error in 1983 (Senders and Moray, 1991). During this workshop, Erik Hollnagel was asked to enlighten the audience on the differences between errors, mistakes, faults and slips. While he tried to shrug off the assignment as "irritating," Hollnagel articulated what Chapanis had pointed out almost four decades earlier: " 'human error' is just one explanation out of several possible for an observed performance." "Human error" is in fact, he said, a label for a presumed cause. If we see something that has gone wrong (the airplane landed belly-up because the gear instead of the flaps was retracted), we may infer that the cause was "human error." This leads to all kinds of scientific trouble. We can hope to make somewhat accurate predictions about outcomes. But causes? By having only outcomes to observe, how can we ever make meaningful predictions about their supposed causes except in the most rigorously deterministic universe (which ours is not)?

The conclusion in 1983 was the need for a better theory of human systems in action, particularly as it relates to the social, organizational, and engineered context in which people do their work. This call echoed William James' functionalism at the turn of the twentieth century, and was taken up by the ecological psychology of Gibson and others after the War (Heft, 1999). What turned out to be more interesting is a good description of the circumstances in which observed problems occur – quite different from searching for supposed "psychological error mechanisms" inside an operator's head. The focus this book is to understand how systematic features of people's environment can reasonably (and predictably) trigger particular actions; actions that make sense given the situation that helped bring them forth. Studying how the system functions as it confronts variations and trouble reveals how safety is created by people in various roles and points to new leverage points for improving safety in complex systems.

The meeting at Bellagio was followed by a workshop in Bad Homburg on new technology and human error in 1986 (Rasmussen, Duncan, and Leplat, 1987), World Bank meetings on safety control and risk management in 1988 and 1989 (e.g., Rasmussen and Batstone, 1989), Reason's elaboration of the latent failure approach (1990; 1997), the debate triggered by Dougherty's editorial in *Reliability Engineering and System Safety* (1990), Hollnagel's *Human Reliability Analysis: Context and Control* (1993) and a series of four workshops sponsored by a US National Academy of Sciences panel from 1990 to 1993 that examined human error from individual, team, organizational, and design perspectives. Between then and today lies a multitude of developments, including the increasing interest in High Reliability Organizations (Rochlin, 1999) and its dialogue with what has become known as Normal Accident Theory (Perrow, 1984), the aftermath of two Space Shuttle accidents, each of which has received extensive public, political, investigatory, and scholarly attention (e.g., Vaughan, 1996; CAIB, 2003), and the emergence of Resilience Engineering (Hollnagel, Woods and Leveson, 2006).

Research in this area is charged. It can never be conducted by disinterested, objective, detached observers. Researchers, like any other people, have certain goals that influence what they see. When the label "human error" becomes the starting point for investigations, rather than a conclusion, the goal of the research must be how to produce change in organizations, in systems, and in technology to increase safety and reduce the risk of disaster. Whether researchers want to recognize it or not, we are participants in the processes of dealing with the aftermath of failure; we are participants in the process of making changes to prevent the failures from happening again.

This means that the label "human error" is inextricably bound up with extra-research issues. The interest in the topic derives from the real world, from the desire to avoid disasters. The potential changes that could be made in real-world hazardous systems to address a "human error problem" inevitably involve high consequences for many stakeholders. Huge investments have been made in technological systems, which cannot be easily changed, because some researcher claims that the incidents relate to design flaws that encourage the possibility of human error. When a researcher asserts that a disaster is due to latent organizational factors and not to the proximal events and actors, he or she is asserting a prerogative to re-design the jobs and responsibilities of hundreds of workers and managers. The factors seen as contributors to a disaster by a researcher could be drawn into legal battles concerning financial liability for the damages and losses associated with an accident, or even, as we have seen recently, criminal liability for operators and managers alike (Dekker, 2007). Laboratory researchers may offer results on biases found in the momentary reasoning of college students while performing artificial tasks. But how much these biases "explain" the human contribution to a disaster is questionable, particularly when the researchers making the claims have not examined the disaster, or the anatomy of disasters and near misses in detail (e.g., Klein, 1989).

FROM ELIMINATING ERROR TO ENHANCING ADAPTIVE CAPACITY

There is an almost irresistible notion that we are custodians of already safe systems that need protection from unreliable, erratic human beings (who get tired, irritable, distracted, do not communicate well, have all kinds of problems with perception, information processing, memory, recall, and much, much more). This notion is unsupported by empirical evidence when one examines how complex systems work. It is also counterproductive by encouraging researchers and consultants and organizations to treat errors as a thing associated with people as a component – the reification fallacy (a kind of over-simplification), treating a set of interacting dynamic processes as if they were a single object.

Eliminating this thing becomes the target of more rigid rules, tighter monitoring of other people, more automation and computer technology all to standardize practices (e.g., "...the elimination of human error is of particular importance in high-risk industries that demand reliability." Krokos and Baker, 2007, p. 175). Ironically, such efforts have unintended consequences that make systems more brittle and hide the sources of resilience that make systems work despite complications, gaps, bottlenecks, goal conflicts, and complexity.

When you go behind the label "human error," you see people and organizations trying to cope with complexity, continually adapting, evolving along with the changing nature of risk in their operations. Such coping with complexity, however, is not easy to see when we make only brief forays into intricate worlds of practice. Particularly when we wield tools to count and tabulate errors, with the aim to declare war on them and make them go away, we all but obliterate the interesting data that is out there for us to discover and learn how the system actually functions. As practitioners confront different evolving situations, they navigate and negotiate the messy details of their practice to bridge gaps and to join together the bits and pieces of their system, creating success as a balance between the multiple conflicting goals and pressures imposed by their organizations. In fact, operators generally do this job so well, that the adaptations and effort glide out of view for outsiders and insiders alike. The only residue left, shimmering on the surface, are the "errors" and incidents to be fished out by those who conduct short, shallow encounters in the form of, for example, safety audits or error counts. Shallow encounters miss how learning and adaptation are ongoing – without these, safety cannot even be maintained in a dynamic and changing organizational setting and environment – yet these adaptations lie mostly out of immediate view, *behind* labels like "human error."

Our experiences in the cross-disciplinary and international discussions convince us that trying to define the term "error" is a bog that quite easily generates unproductive discussions both among researchers and between researchers and the consumers of research (such as regulators, public policy makers, practitioners, and designers). This occurs partly because there is a huge breadth of system, organizational, human performance and human-machine system issues that can become involved in discussions under the rubric of the term "human error." It also occurs because of the increasing complexity of systems in a highly coupled world. The interactional complexity of modern systems means that component-level and single causes are insufficient explanations for failure. Finally, discussions about error are difficult because people tightly hold onto a set of "folk" notions that are generally quite inconsistent with the evidence that has been gathered about erroneous actions and system disasters. Not surprisingly, these folk theories are still prevalent in design, engineering, researcher and sometimes also practitioner communities. Of course, these folk notions themselves arise from the regularities in how we react to failure, but that is what they are: reactions to failure, not explanations of failure.

To get onto productive tracks about how complex systems succeed and fail – the role of technology change and organizational factors – one must directly address the varying perspectives, assumptions, and misconceptions of the different people interested in the topic of human error. It is important to uncover implicit, unexamined assumptions about "human error" and the human contribution to system failures. Making these assumptions explicit and contrasting them with other assumptions and research results can provide the impetus for a continued substantive theoretical debate.

Therefore, the book provides a summary of the assumptions and basic concepts that have emerged from the cross-disciplinary and international discussions and the research that resulted. Our goal is to capture and synthesize some of the results particularly with respect to cognitive factors, the impact of computer technology, and the effect of the hindsight bias on error analysis. While there is no complete consensus among the participants in this work, the overall result is a new look at the human contribution to safety and to risk. This new look continues to be productive generating new results and ideas about how complex systems succeed and fail and about how people in various roles usually create safety.

PART I
AN INTRODUCTION TO THE SECOND STORY

There is a widespread perception of a "human error problem." "Human error" is often cited as a major contributing factor or "cause" of incidents and accidents. Many people accept the term "human error" as one category of potential causes for unsatisfactory activities or outcomes. A belief is that the human element is unreliable, and that solutions to the "human error problem" reside in changing the people or their role in the system.

This book presents the results of an intense examination of the human contribution to safety. It shows that the story of "human error" is remarkably complex. One way to discover this complexity is to make a shift from what we call the "first story," where human error is the cause, to a second, deeper story, in which the normal, predictable actions and assessments (which some call "human error" after the fact) are the product of systematic processes inside of the cognitive, operational and organizational world in which people work. Second stories show that doing things safely – in the course of meeting other goals – is always part of people's operational practice. People, in their different roles, are aware of potential paths to failure, and develop failure sensitive strategies to forestall these possibilities. People are a source of adaptability required to cope with the variation inherent in a field of activity.

Another result of the Second Story is the idea that complex systems have a sharp end and a blunt end. At the sharp end, practitioners directly interact with the hazardous process. At the blunt end, regulators, administrators, economic policy makers, and technology suppliers control the resources, constraints, and multiple incentives and demands that sharp end practitioners must integrate and balance. The story of both success and failure consists of how sharp-end practice adapts to cope with the complexities of the processes they monitor, manage and control, and how the strategies of the people at the sharp end are shaped by the resources and constraints provided by the blunt end of the system.

Failure, then, represents breakdowns in adaptations directed at coping with complexity. Indeed, the enemy of safety is not the human: it is complexity. Stories of how people succeed and sometimes fail in their pursuit of success reveal different sources of complexity as the mischief makers – cognitive, organizational, technological. These sources form an important topic of this book.

This first part of the book offers an overview of these and other results of the deeper study of "human error." It presents 15 premises that recur frequently throughout the book:

1. "Human error" is an attribution after the fact.
2. Erroneous assessments and actions are heterogeneous.
3. Erroneous assessments and actions should be taken as the starting point for an investigation, not an ending.
4. Erroneous actions and assessments are a symptom, not a cause.
5. There is a loose coupling between process and outcome.
6. Knowledge of outcome (hindsight) biases judgments about process.
7. Incidents evolve through the conjunction of several failures/factors.
8. Some of the contributing factors to incidents are always in the system.
9. The same factors govern the expression of expertise and of error.
10. Lawful factors govern the types of erroneous actions or assessments to be expected.
11. Erroneous actions and assessments are context-conditioned.
12. Enhancing error tolerance, error detection, and error recovery together produce safety.
13. Systems fail.
14. Failures involve multiple groups, computers, and people, even at the sharp end.
15. The design of artifacts affects the potential for erroneous actions and paths towards disaster.

The rest of the book explores four main themes that lie behind the label of human error:

o   how systems-thinking is required because there are the multiple factors each necessary but only jointly sufficient to produce accidents in modern systems (Part II);
o   how operating safely at the sharp end depends on cognitive-system factors as situations evolve and cascade – bringing knowledge to bear, shifting mindset in pace with events, and managing goal-conflicts (Part III);
o   how the clumsy use of computer technology can increase the potential for erroneous actions and assessments (Part IV);
o   how what is labeled human error results from social and psychological attribution processes as stakeholders react to failure and how these oversimplifications block learning from accidents and learning before accidents occur (Part V).

1
THE PROBLEM WITH "HUMAN ERROR"

Disasters in complex systems – such as the destruction of the reactor at Three Mile Island, the explosion onboard *Apollo 13*, the destruction of the space shuttles *Challenger* and *Columbia*, the Bhopal chemical plant disaster, the *Herald of Free Enterprise* ferry capsizing, the Clapham Junction railroad disaster, the grounding of the tanker *Exxon Valdez*, crashes of highly computerized aircraft at Bangalore and Strasbourg, the explosion at the Chernobyl reactor, AT&T's Thomas Street outage, as well as more numerous serious incidents which have only captured localized attention – have left many people perplexed. From a narrow, technology-centered point of view, incidents seem more and more to involve mis-operation of otherwise functional engineered systems. Small problems seem to cascade into major incidents. Systems with minor problems are managed into much more severe incidents. What stands out in these cases is the human element.

"Human error" is cited over and over again as a major contributing factor or "cause" of incidents. Most people accept the term human error as one category of potential causes for unsatisfactory activities or outcomes. Human error as a cause of bad outcomes is used in engineering approaches to the reliability of complex systems (probabilistic risk assessment) and is widely cited as a basic category in incident reporting systems in a variety of industries. For example, surveys of anesthetic incidents in the operating room have attributed between 70 and 75 percent of the incidents surveyed to the human element (Cooper, Newbower, and Kitz, 1984; Chopra, Bovill, Spierdijk, and Koornneef, 1992; Wright, Mackenzie, Buchan, Cairns, and Price, 1991). Similar incident surveys in aviation have attributed over 70 percent of incidents to crew error (Boeing, 1993). In general, incident surveys in a variety of industries attribute high percentages of critical events to the category "human error" (see for example, Hollnagel, 1993). The result is the widespread perception of a "human error problem."

One aviation organization concluded that to make progress on safety:

> We must have a better understanding of the so-called human factors which control performance simply because it is these factors which predominate in accident reports. (*Aviation Daily*, November 6, 1992)

The typical belief is that the human element is separate from the system in question and hence, that problems reside either in the human side or in the engineered side of the equation. Incidents attributed to human error then become indicators that the human element is unreliable. This view implies that solutions to a "human error problem" reside in changing the people or their role in the system. To cope with this perceived unreliability of people, the implication is that one should reduce or regiment the human role in managing the potentially hazardous system. In general, this is attempted by enforcing standard practices and work rules, by exiling culprits, by policing of practitioners, and by using automation to shift activity away from people. Note that this view assumes that the overall tasks and system remain the same regardless of the extent of automation, that is the allocation of tasks to people or to machines, and regardless of the pressures managers or regulators place on the practitioners.

For those who accept human error as a potential cause, the answer to the question, what is human error, seems self-evident. Human error is a specific variety of human performance that is so clearly and significantly substandard and flawed when viewed in retrospect that there is no doubt that it should have been viewed by the practitioner as substandard at the time the act was committed or omitted. The judgment that an outcome was due to human error is an attribution that (a) the human performance immediately preceding the incident was unambiguously flawed and (b) the human performance led directly to the negative outcome.

But in practice, things have proved not to be this simple. The label "human error" is very controversial (e.g., Hollnagel, 1993). When precisely does an act or omission constitute an error? How does labeling some act as a human error advance our understanding of why and how complex systems fail? How should we respond to incidents and errors to improve the performance of complex systems? These are not academic or theoretical questions. They are close to the heart of tremendous bureaucratic, professional, and legal conflicts and are tied directly to issues of safety and responsibility. Much hinges on being able to determine how complex systems have failed and on the human contribution to such outcome failures. Even more depends on judgments about what means will prove effective for increasing system reliability, improving human performance, and reducing or eliminating bad outcomes.

Studies in a variety of fields show that the label "human error" is prejudicial and unspecific. It retards rather than advances our understanding of how complex systems fail and the role of human practitioners in both successful and unsuccessful system operations. The investigation of the cognition and behavior of individuals and groups of people, not the attribution of error in itself, points to useful changes for reducing the potential for disaster in large, complex systems. Labeling actions and assessments as "errors" identifies a symptom, not a cause; the symptom should call forth a more in-depth investigation of how a system comprising people, organizations, and technologies both functions and malfunctions (Rasmussen et al., 1987; Reason, 1990; Hollnagel, 1991b; 1993).

Consider this episode which apparently involved a "human error" and which was the stimulus for one of earliest developments in the history of experimental psychology. In 1796 the astronomer Maskelyne fired his assistant Kinnebrook because the latter's observations did not match his own. This incident was one stimulus for another astronomer, Bessel, to examine empirically individual differences in astronomical observations. He found that there were wide differences across observers given the methods of the day and developed what was named the personal equation in an attempt to model and account for these variations (see Boring, 1950). The full history of this episode foreshadows the latest results on human error. The problem was not that one person was the source of errors. Rather, Bessel realized that the standard assumptions about inter-observer accuracies were wrong. The techniques for making observations at this time required a combination of auditory and visual judgments. These judgments were heavily shaped by the tools of the day – pendulum clocks and telescope hairlines – in relation to the demands of the task. In the end, the constructive solution was not dismissing Kinnebrook, but rather searching for better methods for making astronomical observations, re-designing the tools that supported astronomers, and re-designing the tasks to change the demands placed on human judgment.

The results of the recent intense examination of the human contribution to safety and to system failure indicate that the story of "human error" is markedly complex. For example:

- the context in which incidents evolve plays a major role in human performance,
- technology can shape human performance, creating the potential for new forms of error and failure,
- the human performance in question usually involves a set of interacting people,
- the organizational context creates dilemmas and shapes tradeoffs among competing goals,
- the attribution of error after-the-fact is a process of social judgment rather than an objective conclusion.

FIRST AND SECOND STORIES

Sometimes it is more seductive to see human performance as puzzling, as perplexing, rather than as complex. With the rubble of an accident spread before us, we can easily wonder why these people couldn't see what is obvious to us now? After all, all the data was available! Something must be wrong with them. They need re-mediation. Perhaps they need disciplinary action to get them to try harder in the future. Overall, you may feel the need to protect yourself, your system, your organization from these erratic and unreliable other people.

Plus, there is a tantalizing opportunity, what seems like an easy way out – computerize, automate, proceduralize even more stringently – in other words, create a world without those unreliable people who aren't sufficiently careful or motivated. Ask everybody else to try a little harder, and if that still does not work, apply new technology to take over (parts of) their work.

But where you may find yourself puzzled by erratic people, research sees something quite differently. First, it finds that success in complex, safety-critical work depends very much on expert human performance as real systems tend to run degraded, and plans/

algorithms tend to be brittle in the face of complicating factors. Second, the research has discovered many common and predictable patterns in human-machine problem-solving and in cooperative work. There are lawful relationships that govern the different aspects of human performance, cognitive work, coordinated activity and, interestingly, our reactions to failure or the possibility of failure. These are not the natural laws of physiology, aerodynamics, or thermodynamics. They are the control laws of cognitive and social sciences (Woods and Hollnagel, 2006).

The misconceptions and controversies on human error in all kinds of industries are rooted in the collision of two mutually exclusive world views. One view is that erratic people degrade an otherwise safe system. In this view, work on safety means protecting the system (us as managers, regulators and consumers) from unreliable people. We could call this the Ptolemaic world view (the sun goes around the earth). The other world view is that people create safety at all levels of the socio-technical system by learning and adapting to information about how we all can contribute to success and failure. This, then, is a Copernican world view (the earth goes around the sun). Progress comes from helping people create safety. This is what the science says: help people cope with complexity to achieve success. This is the basic lesson from what is now called "New Look" research about error that began in the early 1980s, particularly with one of its founders, Jens Rasmussen.

We can blame and punish under whatever labels are in fashion but that will not change the lawful factors that govern human performance nor will it make the sun go round the earth. So are people sinners or are they saints? This is an old theme, but neither view leads anywhere near to improving safety. This book provides a comprehensive treatment of the Copernican world view or the paradigm that people create safety by coping with varying forms of complexity. It provides a set of concepts about how these processes break down at both the sharp end and the blunt end of hazardous systems.

You will need to shift your paradigm if you want to make real progress on safety in high-risk industries. This shift is, not surprisingly, extraordinarily difficult. Windows of opportunity can be created or expanded, but only if all of us are up to the sacrifices involved in building, extending, and deepening the ways we can help people create safety. The paradigm shift is a shift from a first story, where human error is the cause, to a second, deeper story, in which the normal, predictable actions and assessments which we call "human error" after the fact are the product of systematic processes inside of the cognitive, operational and organizational world in which people are embedded (Cook, Woods and Miller, 1998):

The First Story: Stakeholders claim failure is "caused" by unreliable or erratic performance of individuals working at the sharp end. These sharp-end individuals undermine systems which otherwise work as designed. The search for causes stops when they find the human or group closest to the accident who could have acted differently in a way that would have led to a different outcome. These people are seen as the source or cause of the failure – human error. If erratic people are the cause, then the response is to remove these people from practice, provide remedial training to other practitioners, to urge other practitioners to try harder, and to regiment practice through policies, procedures, and automation.

The Second Story: Researchers, looking more closely at the system in which these practitioners are embedded, reveal the deeper story – a story of multiple contributors that create the conditions that lead to operator errors. Research results reveal systematic factors in both the organization and the technical artifacts that produce the potential for certain kinds of erroneous actions and assessments by people working at the sharp end of the system. In other words, human performance is shaped by systematic factors, and the scientific study of failure is concerned with understanding how these factors lawfully shape the cognition, collaboration, and ultimately the behavior of people in various work domains.

Research has identified some of these systemic regularities that generate conditions ripe with the potential for failure. In particular, we know about how a variety of factors make certain kinds of erroneous actions and assessments predictable (Norman, 1983, 1988). Our ability to predict the timing and number of erroneous actions is very weak, but our ability to foresee vulnerabilities that eventually contribute to failures is often good or very good.

Table 1.1   The contrast between first and second stories of failure

First stories	*Second stories*
Human error (by any other name: violation, complacency) is seen as a cause of failure	Human error is seen as the effect of systemic vulnerabilities deeper inside the organization
Saying what people should have done is a satisfying way to describe failure	Saying what people should have done does not explain why it made sense for them to do what they did
Telling people to be more careful will make the problem go away	Only by constantly seeking out its vulnerabilities can organizations enhance safety

MULTIPLE CONTRIBUTORS AND THE DRIFT TOWARD FAILURE

The research that leads to Second Stories found that doing things safely, in the course of meeting other goals, is always part of operational practice. As people in their different roles are aware of potential paths to failure, they develop failure-sensitive strategies to forestall these possibilities. When failures occurred against this background of usual success, researchers found multiple contributors each necessary but only jointly sufficient and a process of drift toward failure as planned defenses eroded in the face of production pressures and change. These small failures or vulnerabilities are present in the organization or operational system long before an incident is triggered. All complex systems contain such conditions or problems, but only rarely do they combine to create an accident. The research revealed systematic, predictable organizational factors at work, not simply erratic individuals.

This pattern occurs because, in high consequence, complex systems, people recognize the existence of various hazards that threaten to produce accidents. As a result, they develop technical, human, and organizational strategies to forestall these vulnerabilities. For example, people in health care recognize the hazards associated with the need to deliver multiple drugs to multiple patients at unpredictable times in a hospital setting and use computers, labeling methods, patient identification cross-checking, staff training and other methods to defend against misadministration. Accidents in such systems occur when multiple factors together erode, bypass, or break through the multiple defenses creating the trajectory for an accident. While each of these factors is necessary for an accident, they are only jointly sufficient. As a result, there is no single cause for a failure but a dynamic interplay of multiple contributors. The search for a single or root cause retards our ability to understand the interplay of multiple contributors.

Because there are multiple contributors, there are also multiple opportunities to redirect the trajectory away from disaster. An important path to increased safety is enhanced opportunities for people to recognize that a trajectory is heading closer towards a poor outcome and increased opportunities to recover before negative consequences occur. Factors that create the conditions for erroneous actions or assessments, reduce error tolerance, or block error recovery, degrade system performance, and reduce the "resilience" of the system.

SHARP AND BLUNT ENDS OF PRACTICE

The second basic result in the Second Story is to depict complex systems such as health care, aviation and electrical power generation as having a sharp and a blunt end. At the sharp end, practitioners, such as pilots, spacecraft controllers, and, in medicine, as nurses, physicians, technicians, pharmacists, directly interact with the hazardous process. At the blunt end of the system regulators, administrators, economic policy makers, and technology suppliers control the resources, constraints, and multiple incentives and demands that sharp-end practitioners must integrate and balance. As researchers have investigated the Second Story over the last 30 years, they realized that the story of both success and failure is (a) how sharp-end practice adapts to cope with the complexities of the processes they monitor, manage and control and (b) how the strategies of the people at the sharp end are shaped by the resources and constraints provided by the blunt end of the system.

Researchers have studied sharp-end practitioners directly through various kinds of investigations of how they handle different evolving situations. In these studies we see how practitioners cope with the hazards that are inherent in the system. System operations are seldom trouble-free. There are many more opportunities for failure than actual accidents. In the vast majority of cases, groups of practitioners are successful in making the system work productively and safely as they pursue goals and match procedures to situations. However, they do much more than routinely following rules. They also resolve conflicts, anticipate hazards, accommodate variation and change, cope with surprise, work around obstacles, close gaps between plans and real situations, detect and recover from miscommunications and misassessments.

THE PROBLEM WITH "HUMAN ERROR" 9

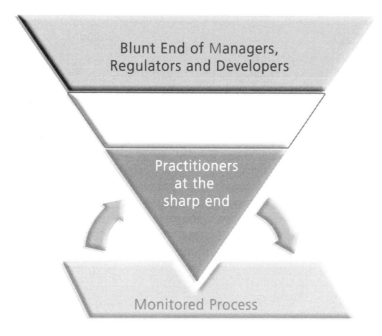

Figure 1.1   The sharp and blunt ends of a complex system

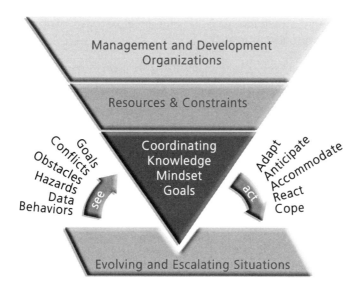

Figure 1.2   At the sharp of a complex system. The interplay of problem-demands and practitioners' expertise at the sharp end govern the expression of expertise and error. The resources available to meet problem demands are provided and constrained by the organizational context at the blunt end of the system

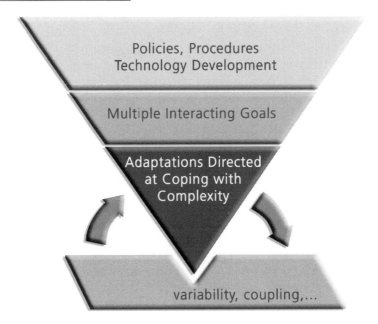

Figure 1.3 The sharp end of a complex system is characterized by how practitioners adapt to cope with complexity. There are different forms of complexity to cope with such, as variability and coupling in the monitored process and multiple interacting goals in the organizational context

ADAPTATIONS DIRECTED AT COPING WITH COMPLEXITY

In their effort after success, practitioners are a source of adaptability required to cope with the variation inherent in the field of activity, for example, complicating factors, surprises, and novel events (Rasmussen, 1986). To achieve their goals in their scope of responsibility, people are always anticipating paths toward failure. Doing things safely is always part of operational practice, and we all develop failure sensitive strategies with the following regularities:

1. People's (groups' and organizations') strategies are sensitive to anticipate the potential paths toward and forms of failure.
2. We are only partially aware of these paths.
3. Since the world is constantly changing, the paths are changing.
4. Our strategies for coping with these potential paths can be weak or mistaken. Updating and calibrating our awareness of the potential paths is essential for avoiding failures.
5. All can be overconfident that they have anticipated all forms, and overconfident that the strategies deployed are effective. As a result, we mistake success as built in rather than the product of effort.

6. Effort after success in a world of changing pressures and potential hazards is fundamental.

In contrast with the view that practitioners are the main source of unreliability in an otherwise successful system, close examination of how the system works in the face of everyday and exceptional demands shows that people in many roles actually "make safety" through their efforts and expertise. People actively contribute to safety by blocking or recovering from potential accident trajectories when they can carry out these roles successfully. To understand episodes of failure you have to first understand usual success – how people in their various roles learn and adapt to create safety in a world fraught with hazards, tradeoffs, and multiple goals.

RESILIENCE

People adapt to cope with complexity (Rasmussen, 1986; Woods, 1988). This notion is based on Ashby's Law of Requisite Variety for control of any complex system. The law is most simply stated as *only variety can destroy variety* (Ashby, 1956). In other words, operational systems must be capable of sufficient variation in their potential range of behavior to match the range of variation that affects the process to be controlled (Hollnagel and Woods, 2005).

The human role at the sharp end is to "make up for holes in designers' work" (Rasmussen, 1981); in other words, to be resilient or robust when events and demands do not fit preconceived and routinized paths (textbook situations). There are many forms of variability and complexity inherent in the processes of transportation, power generation, health care, or space operations. There are goal conflicts, dilemmas, irreducible forms of uncertainty, coupling, escalation, and always a potential for surprise. These forms of complexity can be modeled at different levels of analysis, for example complications that can arise in diagnosing a dynamic system, or in modifying a plan in progress, or those that arise in coordinating multiple actors representing different subgoals and parts of a problem. In the final analysis, the complexities inherent in the processes we manage and the finite resources of all operational systems create tradeoffs (Hollnagel, 2009).

Safety research seeks to identify factors that enhance or undermine practitioners' ability to adapt successfully. In other words, research about safety and failure is about how expertise, in a broad sense distributed over the personnel and artifacts that make up the operational system, is developed and brought to bear to handle the variation and demands of the field of activity.

BLUNT END

When we look at the factors that degrade or enhance the ability of sharp end practice to adapt to cope with complexity we find the marks of organizational factors. Organizations that manage potentially hazardous technical operations remarkably successfully (or high reliability organizations) have surprising characteristics (Rochlin, 1999). Success was not related to how these organizations avoided risks or reduced errors, but rather how these high reliability organizations created safety by anticipating and planning for unexpected events and future surprises. These organizations did not take past success as a reason for confidence. Instead they continued to invest in anticipating the changing potential for failure because of the deeply held understanding that their knowledge base was fragile in the face of the hazards inherent in their work and the changes omnipresent in their environment.

Safety for these organizations was not a commodity, but a value that required continuing reinforcement and investment. The learning activities at the heart of this process depended on open flow of information about the changing face of the potential for failure. The high reliability organizations valued such information flow, used multiple methods to generate this information, and then used this information to guide constructive changes without waiting for accidents to occur. The human role at the blunt end then is to appreciate the changing match of sharp-end practice and demands, anticipating the changing potential paths to failure – assessing and supporting resilience or robustness.

BREAKDOWNS IN ADAPTATION

The theme that leaps out at the heart of these results is that failure represents breakdowns in adaptations directed at coping with complexity (see Woods and Branlat, in press for new results on how adaptive systems fail). Success relates to organizations, groups and individuals who are skillful at recognizing the need to adapt in a changing, variable world and in developing ways to adapt plans to meet these changing conditions despite the risk of negative side effects.

Studies continue to find two basic forms of breakdowns in adaptation (Woods, O'Brien and Hanes, 1987):

o Under-adaptation where rote rule following persisted in the face of events that disrupted ongoing plans and routines.
o Over-adaptation where adaptation to unanticipated conditions was attempted without the complete knowledge or guidance needed to manage resources successfully to meet recovery goals.

In these studies, either local actors failed to adapt plans and procedures to local conditions, often because they failed to understand that the plans might not fit actual circumstances, or they adapted plans and procedures without considering the larger goals and constraints in the situation. In the latter case, the failures to adapt often involved missing side effects of the changes in the replanning process.

SIDE EFFECTS OF CHANGE

Systems exist in a changing world. The environment, organization, economics, capabilities, technology, and regulatory context all change over time. This backdrop of continuous systemic change ensures that hazards and how they are managed are constantly changing. Progress on safety concerns anticipating how these kinds of changes will create new vulnerabilities and paths to failure even as they provide benefits on other scores.

The general lesson is that as capabilities, tools, organizations and economics change, vulnerabilities to failure change as well – some decay, new forms appear. The state of safety in any system is always dynamic, and stakeholder beliefs about safety and hazard also change.

Another reason to focus on change is that systems usually are under severe resource and performance pressures from stakeholders (pressures to become faster, better, and cheaper all at the same time). First, change under these circumstances tends to increase coupling, that is, the interconnections between parts and activities, in order to achieve greater efficiency and productivity. However, research has found that increasing coupling also increases operational complexity and increases the difficulty of the problems practitioners can face. Second, when change is undertaken to improve systems under pressure, the benefits of change may be consumed in the form of increased productivity and efficiency and not in the form of a more resilient, robust and therefore safer system.

It is the complexity of operations that contributes to human performance problems, incidents, and failures. This means that changes, however well-intended, that increase or create new forms of complexity will produce new forms of failure (in addition to other effects). New capabilities and improvements become a new baseline of comparison for potential or actual paths to failure. The public doesn't see the general course of improvement; they see dreadful failure against a background normalized by usual success.

Future success depends on the ability to anticipate and assess how unintended effects of economic, organizational and technological change can produce new systemic vulnerabilities and paths to failure. Learning about the impact of change leads to adaptations to forestall new potential paths to and forms of failure.

COMPLEXITY IS THE OPPONENT

First stories place us in a search to identify a culprit. The enemy becomes other people, those who we decide after-the-fact are not as well intentioned or careful as we are. On the other hand, stories of how people succeed and sometimes fail in their effort after success reveal different forms of complexity as the mischief makers. Yet as we pursue multiple interacting goals in an environment of performance demands and resource pressures these complexities intensify.

Thus, the enemy of safety is complexity. Progress is learning how to tame the complexity that arises from achieving higher levels of capability in the face of resource pressures (Woods, Patterson and Cook, 2006; Woods, 2005). This directs us to look at sources of complexity and changing forms of complexity and leads to the recognition that tradeoffs are at the core of safety (Hollnagel, 2009). Ultimately, Second Stories capture these complexities and lead us to the strategies to tame them. Thus, they are part of the feedback and monitoring process for learning and adapting to the changing pattern of vulnerabilities.

REACTIONS TO FAILURE: THE HINDSIGHT BIAS

The First Story is ultimately barren. The Second Story points the way to constructive learning and change. Why, then, does incident investigation usually stop at the First Story? Why does the attribution of human error seem to constitute a satisfactory explanation for incidents and accidents? There are several factors that lead people to stop with the First Story, but the most important is hindsight bias.

Incidents and accidents challenge stakeholders' belief in the safety of the system and the adequacy of the defenses in place. Incidents and accidents are surprising, shocking events that demand explanation so that stakeholders can resume normal activities in providing or consuming the products or services from that field of practice. As a result, after-the-fact stakeholders look back and make judgments about what led to the accident or incident. This is a process of human judgment where people – lay people, scientists, engineers, managers, and regulators – judge what "caused" the event in question.

In this psychological and social judgment process people isolate one factor from among many contributing factors and label it as the "cause" for the event to be explained. People tend to do this despite the fact there are always several necessary and sufficient conditions for the event. Researchers try to understand the social and psychological factors that lead people to see one of these multiple factors as "causal" while relegating the other necessary conditions to background status. One of the early pieces of work on how people attribute causality is Kelley (1973). A more recent treatment of some of the factors can be found in Hilton (1990). From this perspective, "error" research studies the social and psychological processes which govern our reactions to failure as stakeholders in the system in question.

Our reactions to failure as stakeholders are influenced by many factors. One of the most critical is that, after an accident, we know the outcome. Working backwards from this knowledge, it is clear which assessments or actions were critical to that outcome. It is easy for us with the benefit of hindsight to say, "How could they have missed x?" or "How could they have not realized that x would obviously lead to y?" Fundamentally, this omniscience is not available to any of us before we know the results of our actions.

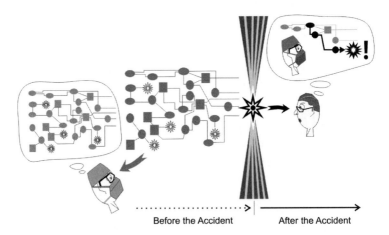

Figure 1.4   Hindsight bias simplifies the situation, difficulties, and hazards faced before outcome is known

Knowledge of outcome biases our judgment about the processes that led up to that outcome. We react, after the fact, as if this knowledge were available to operators. This oversimplifies or trivializes the situation confronting the practitioners, and masks the processes affecting practitioner behavior before-the-fact. Hindsight bias blocks our ability to see the deeper story of systematic factors that predictably shape human performance.

The hindsight bias is one of the most reproduced research findings relevant to accident analysis and reactions to failure. It is the tendency for people to "consistently exaggerate what could have been anticipated in foresight" (Fischhoff, 1975). Studies have consistently shown that people have a tendency to judge the quality of a process by its outcome. Information about outcome biases their evaluation of the process that preceded it. Decisions and actions followed by a negative outcome will be judged more harshly than if the same process had resulted in a neutral or positive outcome.

Research has shown that hindsight bias is very difficult to remove. For example, the bias remains even when those making the judgments have been warned about the phenomenon and advised to guard against it. The First Story of an incident seems satisfactory because knowledge of outcome changes our perspective so fundamentally. One set of experimental studies of the biasing effect of outcome knowledge can be found in Baron and Hershey (1988). Lipshitz (1989) and Caplan, Posner and Cheney (1991) provide demonstrations of the effect with actual practitioners. See Fischhoff (1982) for one study of how difficult it is for people to ignore outcome information in evaluating the quality of decisions.

DEBIASING THE STUDY OF FAILURE

The First Story leaves us with an impoverished view of the factors that shape human performance. In this vacuum, folk models spring up about the human contribution to risk and safety. Human behavior is seen as fundamentally unreliable and erratic (even otherwise effective practitioners occasionally and unpredictably blunder). These folk models, which regard "human error" as the cause of accidents, mislead us. These folk models create an environment where accidents are followed by a search for a culprit and solutions that consist of punishment and exile for the apparent culprit and increased regimentation or remediation for other practitioners as if the cause resided in defects inside people.

These countermeasures are ineffective or even counterproductive because they miss the deeper systematic factors that produced the multiple conditions necessary for failure. Other practitioners, regardless of motivation levels or skill levels, remain vulnerable to the same systematic factors. If the incident sequence included an omission of an isolated act, the memory burdens imposed by task mis-design are still present as a factor ready to undermine execution of the procedure. If a mode error was part of the failure chain, the computer interface design still creates the potential for this type of miscoordination to occur. If a double bind was behind the actions that contributed to the failure, that goal conflict remains to perplex other practitioners.

Getting to Second Stories requires overcoming the hindsight bias. The hindsight bias fundamentally undermines our ability to understand the factors that influenced practitioner behavior. Knowledge of outcome causes reviewers to oversimplify the problem-solving situation practitioners face. The dilemmas, the uncertainties, the tradeoffs, the attentional demands, and double binds faced by practitioners may be missed or under-emphasized when an incident is viewed in hindsight. Typically, hindsight bias makes it seem that participants failed to account for information or conditions that "should have been obvious" or behaved in ways that were inconsistent with the (now known to be) significant information. Possessing knowledge of the outcome, because of the hindsight bias, trivializes the situation confronting the practitioner, who cannot know the outcome before-the-fact, and makes the "correct" choice seem crystal clear.

The difference between everyday or "folk" reactions to failure and investigations of the factors that influence human performance is that researchers use methods designed to remove hindsight bias to see the factors that influenced the behavior of the people in the situation before the outcome is known (Dekker, 2006).

When your investigation stops with the First Story and concludes with the label "human error" (under whatever name), you lose the potential for constructive learning and change. Ultimately, the real hazards to your organization are inherent in the underlying system, and you miss them all if you just tell yourself and your stakeholders a First Story. Yet, independent of what you say, other people and other organizations are acutely aware of many of these basic hazards in their field of practice, and they actively work to devise defenses to guard against them. This effort to make safety is needed continuously. When these efforts break down and we see a failure, you stand to gain a lot of new information, not about the innate fallibilities of people, but about the nature of the threats to your complex systems and the limits of the countermeasures you have put in place.

HUMAN PERFORMANCE: LOCAL RATIONALITY

Why do people do what they do? How could their assessments and actions have made sense to them at the time? The concept of local rationality is critical for studying human performance. No pilot sets out to fly into the ground on today's mission. No physician intends to harm a patient through their actions or lack of intervention. If they do intend this, we speak not of error or failure but of suicide or euthanasia.

After-the-fact, based on knowledge of outcome, outsiders can identify "critical" decisions and actions that, if different, would have averted the negative outcome. Since these "critical" points are so clear to you with the benefit of hindsight, you could be tempted to think they should have been equally clear and obvious to the people involved in the incident. These people's failure to see what is obvious now to you seems inexplicable and therefore irrational or even perverse. In fact, what seems to be irrational behavior in hindsight turns out to be quite reasonable from the point of view of the demands practitioners face and the resources they can bring bear.

Peoples' behavior is consistent with Simon's (1969) principle of bounded rationality – that is, people use their knowledge to pursue their goals. What people do makes sense given their goals, their knowledge and their focus of attention at the time. Human (and machine) problem-solvers possess finite capabilities. There are bounds to the data that they pick up or search out, limits to the knowledge that they possess, bounds to the knowledge that they activate in a particular context, and conflicts among the multiple goals they must achieve. In other words, people's behavior is rational when viewed from the locality of their knowledge, their mindset, and the multiple goals they are trying to balance.

This means that it takes effort (which consumes limited resources) to seek out evidence, to interpret it (as relevant), and to assimilate it with other evidence. Evidence may come in over time, over many noisy channels. The underlying process may yield information only in response to diagnostic interventions. Time pressure, which compels action (or the de facto decision not to act), makes it impossible to wait for all evidence to accrue. Multiple goals may be relevant, not all of which are consistent. It may not be clear, in foresight, which goals are the most important ones to focus on at any one particular moment in time. Human problem-solvers cannot handle all the potentially relevant information, cannot activate and hold in mind all of the relevant knowledge, and cannot entertain all potentially relevant trains of thought. Hence, rationality must be local – attending to only a subset of the possible knowledge, lines of thought, and goals that could be, in principle, relevant to the problem (Simon, 1957; Newell, 1982).

Though human performance is locally rational, after-the-fact we (and the performers themselves) may see how that behavior contributed to a poor outcome. This means that "error" research, in the sense of understanding the factors that influence human performance, needs to explore how limited knowledge (missing knowledge or misconceptions), how a limited and changing mindset, and how multiple interacting goals shape the behavior of the people in evolving situations. All of us use local rationality in our everyday communications. As Bruner (1986, p. 15) put it:

> We characteristically assume that what somebody says must make sense, and we will, when in doubt about what sense it makes, search for or invent an interpretation of the utterance to give it sense.

In other words, this type of error research reconstructs what the view was like (or would have been like) had we stood in the same situation as the participants. If we can understand how the participants' knowledge, mindset, and goals guided their behavior, then we can see how they were vulnerable to breakdown given the demands of the situation they faced. We can see new ways to help practitioners activate relevant knowledge, shift attention among multiple tasks in a rich, changing data field, and recognize and balance competing goals.

2
BASIC PREMISES

❖

Designing human error out of systems was one of the earliest activities of human factors (e.g., Fitts and Jones, 1947). Error counts have been used as a measure of performance in laboratory studies since the beginning of experimental psychology. In fact an episode involving a "human error" was the stimulus for one of the earliest developments in experimental psychology. While error has a long history in human factors and experimental psychology, the decade of the 1980s marked the beginning of an especially energetic period for researchers exploring issues surrounding the label "human error." This international and cross-disciplinary debate on the nature of erroneous actions and assessments has led to a new paradigm about what is error, how to study error, and what kinds of countermeasures will enhance safety. This chapter is an overview of these results. It also serves as an introduction to the later chapters by presenting basic concepts that recur frequently throughout the book.

FIFTEEN PREMISES

The starting point for going behind the label human error is that:

> "Human error" is an attribution after the fact.

Attributing an outcome as the result of error is a judgment about human performance. Such a judgment is rarely applied except when an accident or series of events have occurred that ended with a bad outcome or nearly did so. Thus, these judgments are made ex post facto, with the benefit of hindsight about the outcome or close call. This factor makes it difficult to attribute specific incidents and outcomes to "human error" in a consistent way. Traditionally, error has been seen as a thing in itself – a kind of cause of incidents, a meaningful category that can be used to aggregate specific instances. As a thing, different instances of error can be lumped together and counted as in laboratory studies of human performance or as in risk analyses. Different kinds of errors could be ignored safely and error treated as a homogenous category. In the experimental psychology laboratory, for example, errors are counted as a basic unit of measurement for comparing performance across various factors. This use of error, however, assumes that all types of errors can be combined in a homogenous category, that all specific errors can be treated as equivalent occurrences. This may be true when one has reduced a task to a minimum of content and context as is traditional in laboratory tasks. But real-world, complex tasks carried out by domain practitioners embedded in a larger temporal and organizational context are diverse. The activities and the psychological and behavioral concepts that are involved in these tasks and activities are correspondingly diverse. Hence, the resulting observable erroneous actions and assessments are diverse. In other words, in real fields of practice (where real hazards exist):

> Erroneous assessments and actions are heterogeneous.

One case may involve diagnosis; another may involve perceptual motor skills. One may involve system X and another system Y. One may occur during maintenance, another during operations. One may occur when there are many people interacting; another may occur when only one or a few people are present.

Noting the heterogeneity of errors was one of the fundamental contributions made by John Senders to begin the new and intensive look at human error in 1980. An understanding of erroneous actions and assessments in the real world means that we cannot toss them into a neat causal category labeled "human error." It is fundamental to see that:

> Erroneous assessments and actions should be taken as the starting point for an investigation, not an ending.

This premise is the cornerstone of the paradigm shift for understanding error (Rasmussen, 1986), and much of the material in this book should help to indicate why this premise is so fundamental.

It is common practice for investigators to see errors simply as a specific and flawed piece of human behavior within some particular task. Consider a simple example. Let us assume that practitioners repeatedly confuse two switches, A and B, and inadvertently actuate the wrong one in some circumstances. Then it seems obvious to describe the behavior as a human error where a specific person confused these two switches. This type of interpretation of errors is stuck in describing the episode in terms of the external mode of appearance or the surface manifestation (these two switches were confused), rather than also searching for descriptions in terms of deeper and more general categorizations and underlying mechanisms. For example, this confusion may be an example of a more abstract category such as a slip of action (see Norman, 1981 or Reason and Mycielska, 1982) or a mode error (see Sarter and Woods, 1995, or Part IV).

Hollnagel (1991a, 1993) calls this the difference between the phenotype (the surface appearance) and the genotype of errors (also see the taxonomy of error taxonomies in Rasmussen et al., 1987). Typically, the explicit or implicit typologies of erroneous actions and assessments, such as those used in formal reporting systems categorize errors only on the basis of phenotypes. They do not go beyond the surface characteristics and local context of the particular episode.

As early as Fitts and Jones (1947), researchers were trying to find deeper patterns that cut across the particular. The work of the 1980s has expanded greatly on the repertoire of genotypes that are related to erroneous actions and assessments. In other words, the research has been searching to expand the conceptual and theoretical basis that explains data on system breakdowns involving people. We will lay out several of these in later chapters: ones that are related to cognitive system factors that influence the formation of intentions to act, and ones that are influenced by skillful or clumsy use of computer technology. If we can learn about or discover these underlying patterns, we gain leverage on how to change human-machine systems and about how to anticipate problems prior to a disaster in particular settings.

Thus, in a great deal of the recent work on error, erroneous actions and assessments are treated as the starting point for an investigation, rather than a conclusion to an investigation. The label "error" should be the starting point for investigation of the dynamic interplay of larger system and contextual factors that shaped the evolution of the incident (and other contrasting incidents). The attribution of "human error" is no longer adequate as an explanation for a poor outcome; the label "human error" is not an adequate stopping rule. It is the investigation of factors that influence the cognition and behavior of groups of people, not the attribution of error in itself, that helps us find useful ways to change systems in order to reduce the potential for disaster and to develop higher reliability human-machine systems. In other words, it is more useful from a system design point of view to see that:

> Erroneous actions and assessments are a symptom, not a cause.

There is a great diversity of notions about what "human error" means. The term is problematic, in part, because it is often used in a way that suggests that a meaningful cause has been identified, namely the human. To shed this causal connotation, Hollnagel (1993, p. 29) has proposed the term "erroneous action," which means "an action that fails to produce the expected result or which produces an unwanted consequence." We prefer this term for the same reason.

Another contributor to the diversity of interpretations about human error is confusion between outcome and process. To talk to each other about error we must be very clear about whether we are referring to bad outcomes or a defect in a process for carrying out some activity. We will emphasize the difference between outcome (or performance) failures and defects in the problem-solving process.

Outcome (or performance) failures are defined in terms of a categorical shift in consequences on some performance dimension. They are defined in terms of some potentially observable standard and in terms of the language of the particular field of activity. If we consider military aviation, some examples of outcome failures might include an unfulfilled mission goal, a failure to prevent or mitigate the consequences of some system failure on the aircraft, or a failure to survive the mission. Typically, an outcome failure (or a near miss) provides the impetus for an accident investigation.

Process defects are departures from some standard about how problems should be solved. Generally, the process defect, instantaneously or over time, leads to or increases the risk of some type of outcome failure. Process defects can be defined in terms of a particular field of activity (e.g., failing to verify that all safety systems came on as demanded following a reactor trip in a nuclear power plant) or cognitively in terms of deficiencies in some cognitive or information processing function (e.g., as slips of action, Norman, 1981; fixations or cognitive lockup, De Keyser and Woods, 1990; or vagabonding, Dorner, 1983).

The distinction between outcome and process is important because the relationship between them is not fixed. In other words:

> There is a loose coupling between process and outcome.

This premise is implicit in Abraham Lincoln's vivid statement about process and outcome:

> If the end brings me out all right what is said against me won't amount to anything. If the end brings me out wrong, ten angels swearing I was right would make no difference.

Today's students of decision making echo Lincoln: "Do not judge the quality of a decision by how it turns out. These decisions are inevitably gambles. No one can think of all contingencies or predict consequences with certainty. Good decisions may be followed by bad outcomes" (Fischhoff, 1982, p. 587; cf. also Edwards, 1984). For example, in critical care medicine it is possible that the physician's assessments, plans, and therapeutic responses are "correct" for a trauma victim, and yet the patient outcome may be less than desirable; the patient's injuries may have been too severe or extensive.

Similarly, not all process defects are associated with bad outcomes. Less than expert performance may be insufficient to create a bad outcome by itself; the operation of other factors may be required as well. This is in part the result of successful engineering such as defenses in depth and because opportunities for detection and recovery occur as the incident evolves.

The loose coupling of process and outcome occurs because incidents evolve along a course that is not preset. Further along there may be opportunities to direct the evolution towards successful outcomes, or other events or actions may occur that direct the incident towards negative consequences.

Consider a pilot who makes a mode error which, if nothing is done about it, would lead to disaster within some minutes. It may happen that the pilot notices certain unexpected indications and responds to the situation, which will divert the incident evolution back onto a benign course. The fact that process defects do not always or even frequently lead to bad outcomes makes it very difficult for people or organizations to understand the nature of error, its detection and recovery.

As a result of the loose coupling between process and outcome, we are left with a nagging problem. Defining human error as a form of process defect implies that there exists some criterion or standard against which the performance has been measured and deemed inadequate. However, what standard should be used? Despite many attempts, no one has succeeded in developing a single and simple answer to this question. However, if we are ambiguous about the particular standard adopted to define "error" in particular studies or incidents, then we greatly retard our ability to engage in a constructive and empirically grounded debate about error. All claims about when an action or assessment is erroneous in a process sense must be accompanied by an explicit statement of the standard used for defining departures from good process.

One kind of standard that can be invoked is a normative model of task performance. For many fields of activity where bad outcomes can mean dire consequences, there are no normative models or there are great questions surrounding how to transfer normative models developed for much simpler situations to a more complex field of activity. For example, laboratory-based normative models may ignore the role of time or may assume that cognitive processing is resource-unlimited.

Another possible kind of standard is standard operating practices (e.g., written policies and procedures). However, work analysis has shown that formal practices and policies often depart substantially from the dilemmas, constraints, and tradeoffs present in the actual workplace (e.g., Hirschhorn, 1993). For realistically complex problems there is often no one best method; rather, there is an envelope containing multiple paths each of which can lead to a satisfactory outcome. This suggests the possibility of a third approach for a standard of comparison. One could use an empirical standard that asks: "What would other similar practitioners have thought or done in this situation?" De Keyser and Woods (1990) called these empirically based comparisons neutral practitioner criteria. A simple example occurred in regard to the Strasbourg aircraft crash (Monnier, 1992). Mode error in pilot interaction with cockpit automation seems to have been a contributor to this accident. Following the accident, several people in the aviation industry noted a few precursor incidents or "dress rehearsals" for the crash where similar mode errors had occurred, although the incidents did not evolve as far towards negative consequences. (At least one of these mode errors resulted in an unexpected rapid descent, and the ground proximity warning system alarm alerted the crew who executed a go-around).

Whatever kind of standard is adopted for a particular study:

> Knowledge of outcome (hindsight) biases judgments about process.

People have a tendency to judge the quality of a process by its outcome. The information about outcome biases their evaluation of the process that was followed (Baron and Hershey, 1988). The loose coupling between process and outcome makes it problematic to use outcome information as an indicator for error in a process. (Part V explains the outcome bias and related hindsight bias and discusses their implications for the study of error.)

Studies of disasters have revealed an important common characteristic:

> Incidents evolve through the conjunction of several failures/factors.

Actual accidents develop or evolve through a conjunction of several small failures, both machine and human (Pew et al., 1981; Perrow, 1984; Wagenaar and Groeneweg, 1987; Reason, 1990). This pattern is seen in virtually all of the significant nuclear power plant incidents, including Three Mile Island, Chernobyl, the Brown's Ferry fire, the incidents examined in Pew et al. (1981), the steam generator tube rupture at the Ginna station (Woods, 1982), and others. In the near miss at the Davis-Besse nuclear station (US. N.R.C., NUREG-1154, 1985), there were about 10 machine failures and several erroneous actions that initiated the loss-of-feedwater accident and determined how it evolved.

In the evolution of an incident, there are a series of interactions between the human-machine system and the hazardous process. One acts and the other responds, which, in turn, generates a response from the first, and so forth. Incident evolution points out that there is some initiating event in some human and technical system context, but there is no single clearly identifiable cause of the accident (Rasmussen, 1986; Senders and Moray, 1991). However, several points during the accident evolution can be identified where the evolution can be stopped or redirected away from undesirable outcomes.

Gaba, Maxwell and DeAnda (1987) applied this idea to critical incidents in anesthesia, and Cook, Woods and McDonald (1991a), also working in anesthesia, identified several different patterns of incident evolution. For example, "acute" incidents present themselves all at once, while in "going sour" incidents, there is a slow degradation of the monitored process (see Woods and Sarter, 2000, for going sour patterns in aviation incidents).

One kind of "going sour" incident, which is called decompensation incidents, occurs when an automatic system's responses mask the diagnostic signature produced by a fault (see Woods and Cook, 2006 and Woods and Branlat, in press). As the abnormal influences produced by a fault persist or grow over time, the capacity of automatic systems to counterbalance or compensate becomes exhausted. At some point they fail to counteract and the system collapses or decompensates. The result is a two-phase signature. In phase 1 there is a gradual falling off from desired states over a period of time. Eventually, if the practitioner does not intervene in appropriate and timely ways, phase 2 occurs – a relatively rapid collapse when the capacity of the automatic systems is exceeded or exhausted. During the first phase of a decompensation incident, the gradual nature of the symptoms can make it difficult to distinguish a major challenge, partially compensated for, from a minor disturbance (see National Transportation Safety Board, 1986a). This can lead to a great surprise when the second phase occurs (e.g., some practitioners who miss the signs associated with the first phase may think that the event began with the collapse; Cook, Woods and McDonald, 1991a). The critical difference between a major challenge and a minor disruption is not the symptoms, per se, but rather the force with which they must be resisted. This case illustrates how incidents evolve as a function of the interaction between the nature of the trouble itself and the responses taken to compensate for that trouble.

Some of the contributing factors to incidents are always in the system.

Some of the factors that combine to produce a disaster are latent in the sense that they were present before the incident began. Turner (1978) discusses the incubation of factors prior to the incident itself, and Reason (1990) refers to potential destructive forces that build up in a system in an explicit analogy to resident pathogens in the body. Thus, latent failures refer to problems in a system that produce a negative effect but whose consequences are not revealed or activated until some other enabling condition is met. Examples include failures that make safety systems unable to function properly if called on, such as the error during maintenance that resulted in the emergency feedwater system being unavailable during the Three Mile Island incident (Kemeny Commission, 1979). Latent failures require a trigger, that is, an initiating or enabling event, that activates its effects or consequences. For example, in the Space Shuttle Challenger disaster, the decision to launch in cold weather was the initiating event that activated the consequences of the latent failure – a highly vulnerable booster-rocket seal design. This generalization means that assessment of the potential for disaster should include a search for evidence about latent failures hidden in the system (Reason, 1990).

When error is seen as the starting point for study, when the heterogeneity of errors (their external mode of appearance) is appreciated, and the difference between outcome and process is kept in mind, then it becomes clear that one cannot separate the study of error from the study of normal human behavior and system function. We quickly find that we are not studying error, but rather, human behavior itself, embedded in meaningful contexts. As Rasmussen (1985) states:

> It ... [is] important to realize that the scientific basis for human reliability considerations will not be the study of human error as a separate topic, but the study of normal human behavior in real work situations and the mechanisms involved in adaptation and learning. (p. 1194)

The point is that:

> The same factors govern the expression of expertise and of error.

Jens Rasmussen frequently quotes Ernst Mach (1905, p. 84) to reinforce this point:

> Knowledge and error flow from the same mental sources, only success can tell one from the other.

Furthermore, to study error in real-world situations necessitates studying groups of individuals embedded in a larger system that provides resources and constraints, rather than simply studying private, individual cognition. To study error is to study the function of the system in which practitioners are embedded. Part III covers a variety of cognitive system factors that govern the expression of error and expertise. It also explores some of the demand factors in complex domains and the organizational constraints that both play an important role in the expression of error and expertise.

Underlying all of the previous premises there is a deeper point:

> Lawful factors govern the types of erroneous actions or assessments to be expected.

Errors are not some mysterious product of the fallibility or unpredictability of people; rather errors are regular and predictable consequences of a variety of factors. In some cases we understand a great deal about the factors involved, while in others we currently know less, or it takes more work to find out. This premise is not only useful in improving a particular system, but also assists in defining general patterns that cut across particular circumstances. Finding these regularities requires examination of the contextual factors surrounding the specific behavior that is judged faulty or erroneous. In other words:

> Erroneous actions and assessments are context-conditioned.

Many kinds of contextual factors are important to human cognition and behavior (see Figures 1.1, 1.2, and 1.3). The demands imposed by the kinds of problems that can occur are one such factor. The constraints and resources imposed by organizational factors are another. The temporal context defined by how an incident evolves is yet another (e.g., from a practitioner's perspective, a small leak that gradually grows into a break is very different from an incident where the break occurs quite quickly). Part III discusses these and many other cognitive system factors that affect the expression of expertise and error.

Variability in behavior and performance turns out to be crucial for learning and adaptation. In some domains, such as control theory, an error signal, as a difference from a target, is informative because it provides feedback about goal achievement and indicates when adjustments should be made. Error, as part of a continuing feedback and improvement process, is information to shape future behavior. However, in certain contexts this variability can have negative consequences. As Rasmussen (1986) puts it, in "unkind work environments" variability becomes an "unsuccessful experiment with unacceptable consequences." This view emphasizes the following important notion:

> Enhancing error tolerance, error detection, and error recovery together produce safety.

Again, according to Rasmussen (1985):

> The ultimate error frequency largely depends upon the features of the work interface which support immediate error recovery, which in turn depends on the observability and reversibility of the emerging unacceptable effects. The feature of reversibility largely depends upon the dynamics and linearity of the system properties, whereas observability depends on the properties of the task interface which will be dramatically influenced by the modern information technology.

Figure 2.1 illustrates the relationship between recovery from error and the negative consequences of error (outcome failures) when an erroneous action or assessment occurs in some hypothetical system. The erroneous action or assessment is followed by a recovery interval, that is, a period of time during which actions can be taken to reverse the effects of the erroneous action or during which no consequences result from the erroneous assessment. If error detection occurs, the assessment is updated or the previous actions are corrected or compensated far before any negative consequences accrue. If not, then an outcome failure has occurred. There may be further recovery intervals during which other outcome consequences (of a more severe nature) may be avoided if detection and recovery actions occur. Note that this schematic – seeing the build up to an accident as

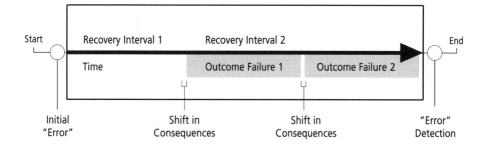

Figure 2.1 The relationship between error recovery and outcome failure. Outcome failures of various types (usually of increasing severity) may be averted if recovery occurs within a particular time span, the length of which depends on the system characteristics

a series of opportunities to detect and revise that went astray – provides one frame for avoiding hindsight bias in analyzing cognitive processes leading up to a failure (e.g., the analyses of foam debris risk prior to the *Columbia* space shuttle launch; *Columbia* Accident Investigation Board, 2003; Woods, 2005).

A field of activity is tolerant of erroneous actions and assessments to the degree that such errors do not immediately or irreversibly lead to negative consequences. An error-tolerant system has a relatively long recovery interval, that is, there are extensive opportunities for reversibility of actions. Error recovery depends on the observability of the monitored process which is in large part a property of the human-computer interface for computerized systems. For example, is it easy to see if there is a mismatch between expected state and the actual state of the system? Several studies show that many human-computer interfaces provide limited observability, that is, they do not provide effective visualization of events, change and anomalies in the monitored process (e.g., Moll van Charante, Cook, Woods, Yue, and Howie, 1993, for automated operating room devices; Woods, Potter, Johannesen, and Holloway, 1991, for intelligent systems for fault management of space vehicle systems; Sarter and Woods, 1993, for cockpit automation). The opaque nature of the interfaces associated with new technology is particularly troubling because it degrades error recovery. Moll van Charante et al. (1993) and Cook, Woods, and Howie (1992) contain data directly linking low observability through the computer interface to critical incidents in the case of one automated operating room device, and Sarter and Woods (1997) link low observability through the interface to problems in mode awareness for cockpit automation (cf. also the Therac-25 accidents, in which a radiation therapy machine delivered massive doses of radiation, for another example where low observability through the computer interface to an automatic system blocked error or failure detection and recovery; Leveson and Turner, 1992).

While design to minimize or prevent erroneous actions is good practice, one cannot eliminate the possibility for error. It seems that the path to high reliability systems critically depends on design to enhance error recovery prior to negative consequences (Lewis and Norman, 1986; Rasmussen, 1986; Reason, 1990). Rasmussen (1985) points out that reported frequencies of "human error" in incident reports are actually counts of errors that were not detected and recovered from, prior to some negative consequence or some criterion for cataloging incidents. Opportunities for the detection and correction of error, and hence tools that support people in doing so are critical influences on how incidents will evolve (see Seifert and Hutchins, 1992, for just one example).

Enhancing error tolerance and error recovery is a common prescription for designing systems (e.g., Norman, 1988). Some methods include:

a. design to prevent an erroneous action, for example, forcing functions which constrain a sequence of user actions along particular paths;
b. design to increase the tolerance of the underlying process to erroneous actions; and
c. design to enhance recovery from errors and failures through effective feedback and visualizations of system function – enhanced observability of the monitored process.

Let us pause and summarize a few important points: failures involve multiple contributing factors. The label error is often used in a way that simply restates the fact that the outcome was undesirable. Error is a symptom indicating the need to investigate the larger operational system and the organizational context in which it functions. In other words:

> Systems fail.

If we examine actual accidents, we will typically find that several groups of people were involved. For example, in the Dallas windshear aircraft crash (National Transportation Safety Board, 1986b), the incident evolution involved the crew of the aircraft in question, what other planes were doing, air traffic controllers, the weather service, company dispatch, company and industry pressures about schedule delays.

> Failures involve multiple groups, computers, and people, even at the sharp end.

One also finds in complex domains that error detection and recovery are inherently distributed over multiple people and groups and over human and machine agents. This is the case in aircraft-carrier flight operations (Rochlin, La Porte, and Roberts, 1987), maritime navigation (Hutchins, 1990), power plant startup (Roth and Woods, 1988) and medication administration (Patterson, Cook, Woods and Render, 2004). Woods et al. (1987) synthesized results across several studies of simulated and actual nuclear power plant emergencies and found that detection and correction of erroneous state assessments came primarily from other crew members who brought a fresh point of view into the situation. Miscommunications between air traffic control and commercial airline flight decks occur frequently, but the air transport system has evolved robust cross-people mechanisms to detect and recover from communication breakdowns, for example, crew cross-checks and read-backs, although miscommunications still can play a role in accidents (National Transportation Safety Board, 1991). Systems for cross-checking occur in pilots' coordination with cockpit automation. For example, pilots develop and are taught cross-check strategies to detect and correct errors that might occur in giving instructions to the flight computers and automation. There is evidence, though, that the current systems are only partially successful and that there is great need to improve the coordination between people and automated agents in error or failure detection (e.g., Sarter and Woods, 1997; Branlat, Anders, Woods and Patterson, 2008).

Systems are always made up of people in various roles and relationships. The systems exist for human purposes. So when systems fail, of course human failure can be found in the rubble. But progress towards safety can be made by understanding the system of people and the resources that they have evolved and their adaptations to the demands of the environment. Thus, when we start at "human error" and begin to investigate the factors that lead to behavior that is so labeled, we quickly progress to studying systems of people embedded in a larger organizational context. In this book we will tend to focus on the sharp-end system, that is, the set of practitioners operating near the process and hazards, the demands they confront, and the resources and constraints imposed by organizational factors.

The perception that there is a "human error problem" is one force that leads to computerization and increased automation in operational systems. As new information and automation technology is introduced into a field of practice what happens to "human error"? The way in which technological possibilities are used in a field of practice affects the potential for different kinds of erroneous actions and assessments. It can reduce the chances for some kinds of erroneous actions or assessments, but it may create or increase the potential for others. In other words:

> The design of artifacts affects the potential for erroneous actions and paths towards disaster.

Artifacts are simply human-made objects. In this context we are interested particularly in computer-based artifacts from individual microprocessor-based devices such as infusion pumps for use in medicine to the suite of automated systems and associated human-computer interfaces present in advanced cockpits on commercial jets. One goal for this book is to focus on the role of design of computer-based artifacts in safety.

Properties of specific computer-based devices or aspects of more general "vectors" of technology change influence the cognition and activities of those people who use them. As a result, technology change can have profound repercussions on system operation, particularly in terms of the types of "errors" that occur and the potential for failure. It is important to understand how technology change shapes human cognition and action in order to see how design can create latent failures which may contribute, given the presence of other factors, to disaster. For example, a particular technology change may increase the coupling in a system (Perrow, 1984). Increased coupling increases the cognitive demands on practitioners. If the computer-based artifacts used by practitioners exhibit "classic" flaws such as weak feedback about system state (what we will term low observability), the combination can function as a latent failure awaiting the right circumstances and triggering events to lead the system close to disaster (see Moll van Charante et al., 1993, for one example of just this sequence of events).

One particular type of technology change, namely increased automation, is assumed by many to be the prescription of choice to cure an organization's "human error problem." One recent example of this attitude comes from a commentary about cockpit developments envisioned for a new military aircraft in Europe:

> The sensing, processing and presentation of such unprecedented quantities of data to inform and protect one man requires new levels of.. system integration. When proved in military service, these automation advances will read directly across to civil aerospace safety. They will also assist the industrial and transport communities' efforts to eliminate 'man-machine interface' disasters like King's Cross, Herald of Free Enterprise, Clapham Junction and Chernobyl. (*Aerospace*, November, 1992, p. 10)

If incidents are the result of "human error," then it seems justified to respond by retreating further into the philosophy that "just a little more technology will be enough" (Woods, 1990b; Billings, 1991). Such a technology-centered approach is more likely to increase the machine's role in the cognitive system in ways that will squeeze the human's role (creating a vicious cycle as evidence of system problems will pop up as more human error). As S. S. Stevens noted (1946):

> The faster the engineers and the inventors served up their 'automatic' gadgets to eliminate the human factor the tighter the squeeze became on the powers of the operator.

And as Norbert Wiener noted some years later (1964, p. 63):

> The gadget-minded people often have the illusion that a highly automatized world will make smaller claims on human ingenuity than does the present one. ... This is palpably false.

Failures to understand the reverberations of technological change on the operational system hinder the understanding of important issues such as what makes problems difficult, how breakdowns occur, and why experts perform well.

Our strategy is to focus on how technology change can increase or decrease the potential for different types of erroneous actions and assessments. Later in the book, we will lay out a broad framework that establishes three inter-related linkages: the effect of technology on the cognitive activities of practitioners; how this, in turn, is linked to the potential for erroneous actions and assessments; and how these can contribute to the potential for disaster.

The concept that the design of the human-machine system, defined very broadly, affects or "modulates" the potential for erroneous actions and assessments, was present at the origins of Human Factors when the presence of repeated "human errors" was treated as a signal pointing to context-specific flaws in the design of human-machine systems (e.g., cockpit control layout). This idea has been reinforced more recently when researchers have identified kinds of design problems in computer-based systems that cut across specific contexts. In general, "clumsy" use of technological powers can create additional mental burdens or other constraints on human cognition and behavior that create opportunities for erroneous actions and assessments by people, especially in high criticality, high workload, high tempo operations (Wiener, 1989; Sarter and Woods, 1997; Woods and Hollnagel, 2006).

Computer-based devices, as typically designed, tend to exhibit classic human-computer cooperation flaws such as lack of feedback on device state and behavior (e.g., Norman, 1990b; Woods, 1995a). Furthermore, these human-computer interactions (HCI) flaws increase the potential for erroneous actions and for erroneous assessments of device state and behavior. The low observability supported by these interfaces and the associated potential for erroneous state assessment is especially troublesome because it impairs the user's ability to detect and recover from failures, repair communication breakdowns, and detect erroneous actions.

These data, along with critical incident studies, directly implicate the increased potential for erroneous actions and the decreased ability to detect errors and failures as one kind of important contributor to actual incidents. The increased potential for error that emanates from poor human-computer cooperation is one type of problem that can be activated and progress towards disaster when in the presence of other potential factors.

Our goals are to expose various design "errors" in human-computer systems that create latent failures, show how devices with these characteristics shape practitioner cognition and behavior, and how these characteristics can create new possibilities for error and new paths to disaster. In addition, we will examine data on how practitioners cope with the complexities introduced by the clumsy use of technological possibilities and how this adaptation process can obscure the role of design and cognitive system factors in incident evolution. This information should help developers detect, anticipate, and recover from designer errors in the development of computerized devices.

HOW COMPLEX SYSTEMS FAIL

Our understanding of how accidents happen has undergone significant changes over the last century (Hollnagel, 2004). Beginning with ideas on industrial safety improvements and the need to contain the risk of uncontrolled energy releases, accidents were initially viewed as the conclusion of a sequence of events (which involved "human errors" as causes or contributors). This has now been replaced by a systemic view in which accidents emerge from the coupling and interdependence of modern systems. The key theme is how system change and evolution produce complexities which challenge people's ability to understand and manage risk in interdependent processes. Inspired in part by new sciences that study complex co-adaptive processes (results on emergent properties, fundamental tradeoffs, non-linear feedback loops, distributed control architectures, and multi-agent simulations), research on safety has shifted from linear cause-effect analyses and reductive models that focus on component level interventions.

Today, safety research focuses more on the ability of systems to recognize, adapt to and absorb disruptions and disturbances, even those that fall beyond the capabilities that the system was trained or designed for. The latest intellectual turn, made by what is called Resilience Engineering, sees how practitioners and organizations, as adaptive, living systems, continually assess and revise their approaches to work in an attempt to balance tradeoffs across multiple goals while remaining sensitive to the possibility of failure. The research studies how safety is created, how organizations learn prior to accidents, how organizations monitor boundaries of safe operation while under pressure to be faster, better and cheaper (Hollnagel, Woods and Leveson, 2006).

COGNITIVE SYSTEMS

The demands that large, complex systems operations place on human performance are mostly cognitive. The third part of the book focuses on cognitive system factors related to the expression of expertise and error. The difference between expert and inexpert human performance is shaped, in part, by three classes of cognitive factors: knowledge factors – how knowledge is brought to bear in specific situations, attentional dynamics – how mindset is formed, focuses and shifts focus as situations evolve and new events occur, and strategic factors – how conflicts between goals are expressed in situations and how these conflicts are resolved. However, these cognitive factors do not apply just to an individual but also to teams of practitioners. In addition, the larger organization context – the blunt end of the system – places constraints and provides resources that shape how practitioners can meet the demands of a specific field of practice.

One of the basic themes that have emerged in more recent work on expertise and error is the need to model team and organizational factors (Hutchins, 1995a). Part III integrates individual, team, and organizational perspectives by viewing operational systems as distributed and joint human-machine cognitive systems. It also lays out the cognitive processes carried out across a distributed system that govern the expression of expertise as well as error in real systems. It explores some of the ways that these processes go off track or break down and increase the vulnerability to erroneous actions.

COMPUTERS

The fourth part of the book addresses the clumsy use of new technological possibilities in the design of computer-based devices and shows how these "design errors" can create the potential for erroneous actions and assessments. Some of the questions addressed in this part include:

- What are these classic design "errors" in human-computer systems, computer-based advisors, and automated systems?
- Why do we see them so frequently in so many settings?
- How do devices with these characteristics shape practitioner cognition and behavior?
- How do practitioners cope with the complexities introduced by clumsy use of technological possibilities?
- What do these factors imply about the human contribution to risk and to safety?

We will refer frequently to mode error as an exemplar of the issues surrounding the impact of computer technology and error. We use this topic as an example extensively because it is an error form that exists only at the intersection of people and technology. Mode error requires a device where the same action or indication means different things in different contexts (i.e., modes) and a person who loses track of the current context. However there is a second and perhaps more important reason that we have chosen this error form as a central exemplar. If we as a community of researchers cannot get design and development organizations to acknowledge, deal with, reduce, and better cope with the proliferation of complex modes, then it will prove difficult to shift design resources and priorities to include a user-centered point of view.

HINDSIGHT

The fourth part of the book examines how the hindsight bias affects the ability to learn from accidents and to learn about risks before accidents occur. It shows how attributions of error are a social and psychological judgment process that occurs as stakeholders struggle to come to grips with the consequences of failures. Many factors contribute to incidents and disasters. Processes of casual attribution influence which of these many factors we focus on and identify as causal. Causal attribution depends on who we are communicating to, on the assumed contrast cases or causal background for that exchange, on the purposes of the inquiry, and on knowledge of the outcome (Tasca, 1990).

Hindsight bias, as indicated above, is the tendency for people to "consistently exaggerate what could have been anticipated in foresight" (Fischhoff, 1975). Studies have consistently shown that people have a tendency to judge the quality of a process by its outcome. The information about outcome biases their evaluation of the process that was followed. Decisions and actions followed by a negative outcome will be judged more harshly than if the same decisions had resulted in a neutral or positive outcome. Indeed this effect is present even when those making the judgments have been warned about the phenomenon and been advised to guard against it (Fischhoff, 1975, 1982).

The hindsight bias leads us to construct "a map that shows only those forks in the road that we decided to take," where we see "the view from one side of a fork in the road, looking back" (Lubar, 1993, p. 1168). Given knowledge of outcome, reviewers will tend to simplify the problem-solving situation that was actually faced by the practitioner. The dilemmas, the uncertainties, the tradeoffs, the attentional demands, and double binds faced by practitioners may be missed or under-emphasized when an incident is viewed in hindsight. Typically, the hindsight bias makes it seem that participants failed to account for information or conditions that "should have been obvious" or behaved in ways that were inconsistent with the (now known to be) significant information. Possessing knowledge of the outcome, because of the hindsight bias, trivializes the situation confronting the practitioner and makes the "correct" choice seem crystal clear.

The hindsight bias has strong implications for studying erroneous actions and assessments and for learning from system failures. If we recognize the role of hindsight and psychological processes of causal judgment in attributing error after-the-fact, then we can begin to devise new ways to study and learn from erroneous actions and assessments and from system failure.

In many ways, the topics addressed in each chapter interact and depend on the concepts introduced in the discussion of other topics from other chapters. For example, the chapter on the clumsy use of computer technology in some ways depends on knowledge of cognitive system factors, but in other ways it helps to motivate the cognitive system framework. There is no requirement to move linearly from one chapter to another. Jump around as your interests and goals suggest.

CAVEATS

There are several topics the book does not address. First, we will not consider research results on how human action sequences can break down including slips of action such as substitution or capture errors. Good reviews are available (Norman, 1981; Reason and Mycielska, 1982; and see Byrne and Bovair, 1997). Second, we do not address the role of fatigue in human performance. Research on fatigue has become important in health care and patient safety (for recent results see Gaba and Howard, 2002). Third, we will not be concerned with work that goes under the heading of Human Reliability Analysis (HRA), because (a) HRA has been dominated by the assumptions made for risk analysis of purely technological systems, assumptions that do not apply to people and human-machine systems very well, and (b) excellent re-examinations of human reliability from the perspective of the new look behind error are available (cf., Hollnagel, 1993, 2004).

PART II
COMPLEX SYSTEMS FAILURE

In the study of accidents, it is important to understand the dynamics and evolution of the conditions that give rise to system breakdowns. Various stakeholders often imagine that the typical path to disaster is a single and major failure of a system component – very often the human. Hence, "human error" often is seen as a cause, or important contributor, to accidents. Studies of the anatomy of disasters in highly technological systems, however, show a different pattern. That which we label "human error" after the fact is never the *cause* of an accident. Rather, it is the cumulative effect of multiple cognitive, collaborative, and organizational factors. This, indeed, is the whole point of the second story: go behind the label "human error" to find out about the systemic factors that gave rise to the behavior in question.

Our understanding of how accidents happen has undergone a dramatic development over the last century (Hollnagel, 2004). Accidents were initially viewed as the conclusion of a sequence of events (which involved "human errors" as causes or contributors). This has now been replaced by a systemic view in which accidents emerge from the complexity of people's activities in an organizational and technical context. These activities are typically focused on preventing accidents, but also involve other goals (throughput, efficiency, cost control) which means that goal conflicts can arise, always under the pressure of limited resources (e.g., time, money, expertise). Accidents emerge from a confluence of conditions and occurrences that are usually associated with the pursuit of success, but in this combination serves to trigger failure instead. Accidents in modern systems arise from multiple contributors each necessary but only jointly sufficient.

In the systemic view, "human errors" are labels for normal, often predictable, assessments and actions that make sense given the knowledge, goals, and focus of attention of people at the time; assessments and actions that make sense in the operational and organizational context that helped bring them forth. "Human error," in other words, is the product of factors that lie deeper inside the operation and organization; and that lie deeper in history too. In the next two chapters we review a number of models of how accidents occur, roughly in historical order, and assess each for their role in helping us understand human error as an effect, rather than a cause. Each model proposes slightly different ideas about what we can find behind the label and what we should do about it. The different models also set the stage for the discussion in Part III on operating at the sharp end by highlighting different constraints and difficulties which challenge operations at the sharp end. Each model emphasizes different aspects about what practitioners need to do to create successful outcomes under goal conflicts and resource pressures.

FROM STOPPING LINEAR SEQUENCES TO THE CONTROL OF COMPLEXITY

Accidents seem unique and diverse, yet safety research identifies systematic features and common patterns. Based on different assumptions about how accidents happen, we have divided our treatment of complex system failure into two chapters. In Chapter 3 we cover models that essentially treat accidents as the outcome of a series of events along a linear pathway, and that see risk as the uncontrolled release of energy:

1. The sequence-of-events model
2. Man-made disaster theory
3. The latent failure model.

Chapter 4 covers the shift to models that examine how accidents emerge from the interaction of a multitude of events, processes and relationships in a complex system, and that take a more interactive, sociological perspective on risk:

1. Normal-accidents theory
2. Control theory
3. High-reliability theory.

Chapter 5 deals with Resilience Engineering, a departure from conventional risk management approaches that build up from assessments of components (e.g., error tabulation, violations, calculation of failure probabilities) and that looks for ways to enhance the ability of organizations to monitor and revise risk models, to create processes that are robust yet flexible, and to use resources proactively in the face of disruptions or ongoing production and economic pressures. According to Resilience Engineering, accidents do not represent a breakdown or malfunctioning of normal system functions, but rather represent the breakdowns in the adaptations necessary to cope with complexity.

Of course this separation into different chapters is not entirely clean, as elements from each set of ideas get borrowed and transferred between approaches.

The first group – linear and latent – have adopted their basic ideas from industrial safety improvements the first half of the twentieth century. Consistent with this lineage, they suggest we think of risk in terms of energy – for example, a dangerous build-up of energy, unintended transfers, or uncontrolled releases of energy (Rosness, Guttormsen, Steiro, Tinmannsvik and Herrera, 2004). This risk needs to be contained, and the most popular way is through a system of barriers: multiple layers whose function it is to stop or inhibit propagations of dangerous and unintended energy transfers. This separates the object-to-be-protected from the source of hazard by a series of defenses (which is a basic notion in the latent-failure model). Other countermeasures include preventing or improving the recognition of the gradual build-up of dangerous energy (something that inspired Man-made disaster theory), reduce the amount of energy (e.g., reduce vehicle speeds or the available dosage of a particular drug in its packaging), prevent the uncontrolled release of energy or safely distribute its release. Such models are firmly rooted in Newtonian visions of cause, particularly the symmetry between cause and effect. Newton's third law of motion is taken as self-evidently applicable: each cause has an equally large effect (which is in turn embedded in the idea of preservation of energy: energy can never disappear out of the universe, only change form). Yet such presumed symmetry (or cause-consequence equivalence) can mislead human error research and practitioners into believing that really bad consequences (a large accident) must have really large causes (very bad or egregious human errors and violations).

The conceptualization of risk as energy to be contained or managed has its roots in efforts to understand and control the physical (or purely technical) nature of accidents. This also spells out the limits of such conceptualization: it is not well-suited to explain the organizational and socio-technical factors behind system breakdown, nor equipped with a language that can meaningfully handle processes of gradual adaptation, risk management, and decision making. The central analogy used for understanding how systems work is the machine, and the chief strategy reductionism. To understand how something works, these models have typically dismantled it and looked at the parts that make up the whole. This approach assumes that we can derive the macro properties of a system (e.g., safety) as a straightforward combination or aggregation of the performance of the lower-order components or subsystems that constitute it. Indeed, the assumption is that safety can be increased by guaranteeing the reliability of the individual system components and the layers of defense against component failure so that accidents will not occur.

The accelerating pace of technological change has introduced more unknowns into our safety-critical systems, and made them increasingly complex. Computer technology, and software in particular, has changed the nature of system breakdowns (see the A330 test flight crash, AWST, 1995; the Ariane 501 failure, Lions, 1996; the GlobalHawk unmanned aerial vehicle accident in 1999, USAF, 1999; and technology assisted friendly fire cases, Loeb, 2002). Accidents can emerge from the complex, non-linear interaction between many reliably operating sub-components. The loss of NASA's Mars Polar Lander, for instance, could be linked to spurious computer signals when the landing legs were deployed during descent towards the Martian surface. This "noise" was normal; it was expected. The onboard software, however, interpreted it as an indication that the craft had landed (which the software engineers were told it would indicate) and shut down the engines prematurely. This caused the spacecraft to crash into the Mars surface. The landing leg extension and software all performed correctly (as specified in their requirements), but the accident emerged from unanticipated interactions between leg-deployment and descent-engine control software (Leveson, 2002; Stephenson et al. 2000).

Accidents where no physical breakage can be found, of course, heighten suspicions about human error. Given that no components in the engineered system malfunctioned or broke, the fault must lie with the people operating the system; with the human component, the human factor. This is indeed is what models in the next chapter tend to do: failures of risk management can get attributed to deficient supervision, ineffective leadership, or lack of appropriate rules and procedures (which points to components that were broken somewhere in the organization). But this just extends reductive thinking, while still searching for broken components to target for intervention.

More recent accident models attempt to make a break from mechanistic, component-oriented images of organizations and risk containment. Instead, they view systems as a whole – a socio-technical system, a co-adaptive or a distributed human-machine cognitive system – and examine the role of emergent properties of these systems – for example, coupling or interdependencies across parts, cascades of disturbances, ability to make cross-checks between groups, or the brittleness or resilience of control and management systems in the face of surprises.

These models try to understand how failure emerges from the normal behaviors of a complex, non-linear system. Emergence means that simple entities, because of their interaction, cross-adaptation and cumulative change, can produce far more complex behaviors as a collective, and produce effects across scales. One common experience is that small changes (e.g., reusing software code in a new-generation system) can lead to huge consequences (enormous releases of energy, such as a Mars Polar Lander crashing onto the surface, or huge overdoses of radioactive energy in cancer treatment with radiation therapy (Leveson, 2002; Johnston, 2006; Cook et al, 2008). Such effects are impossible to capture with linear or sequential models that make Newtonian cause-effect assumptions and cannot accommodate non-linear feedback loops or growth and adaptation. Instead, it takes, for example, complexity theory to understand how simple things can generate very complex outcomes that could not be anticipated by just looking at the parts themselves.

Inspired by recent developments in the study of complexity and adaptation, Resilience Engineering no longer talks about human error at all; instead, it sees safety as something positive, as the *presence* of something. Resilience Engineering focuses on the ability of systems to recognize, adapt to and absorb disruptions and disturbances, especially those that challenge the base capabilities of the system. This concern with adaptation as a central capability that allows living systems to survive in a changing world has been inspired by a number of fields external to the traditional purview of human error research, for example biology, materials science, and physics.

Practitioners and organizations, as adaptive, living systems, continually assess and revise their approaches to work in an attempt to remain sensitive to the possibility of failure. Efforts to create safety, in other words, are ongoing. Strategies that practitioners and organizations (including regulators and inspectors) maintain for coping with potential pathways to failure can be either strong and resilient or weak and brittle. Organizations and people can also become overconfident or mis-calibrate, thinking their strategies are more effective than they really are. High-reliability organizations remain alert for signs that circumstances exist, or are developing, which challenge previously successful strategies (Rochlin, 1993; Gras, Moricot, Poirot-Delpech, and Scardigli, 1994). By knowing and monitoring their boundaries, learning organizations can avoid narrow interpretations of risk and stale strategies.

The principles of organization in a living adaptive system are unlike those of machines. Machines tend to be brittle while living systems gracefully degrade under most circumstances. In the extreme case adaptive systems can respond to very novel conditions such as United Flight 232 in July 1989. After losing control of the aircraft's control surfaces as a result of a center engine failure that ripped fragments through all three hydraulic lines nearby, the crew figured out how to maneuver the aircraft with differential thrust on two remaining engines. They managed to put the crippled DC-10 down at Sioux City, saving 185 lives out of 293.

The systems perspective, and the analogy to living organizations whose stability is dynamically emergent rather than structurally inherent, means that safety is something a system does, not something a system has (Hollnagel, Woods and Leveson, 2006). Failures represent breakdowns in adaptations directed at coping with complexity (Woods, 2003).

3
LINEAR AND LATENT FAILURE MODELS

THE SEQUENCE-OF-EVENTS MODEL

Accidents can be seen as the outcome of a sequence, or chain, of events. This simple, linear way of conceptualizing how events interact to produce a mishap was first articulated by Heinrich in 1931 and is still commonplace today. According to this model, events preceding the accident happen linearly, in a fixed order, and the accident itself is the last event in the sequence. It has been known too as the domino model, for its depiction of an accident as the endpoint in a string of falling dominoes (Hollnagel, 2004). Consistent with the idea of a linear chain of events is the notion of a root cause – a trigger at the beginning of the chain that sets everything in motion (the first domino that falls and then, one by one, the rest). The sequence-of-events idea is pervasive, even if multiple parallel or converging sequences are sometimes depicted to try to capture some of the greater complexity of the precursors to an accident. The idea forms the basic premise in many risk analysis methods and tools such as fault-tree analysis, probabilistic risk assessment, critical path models and more.

Also consistent with a chain of events is the notion of barriers – a separation between the source of hazard and the object or activity that needs protection. Barriers can be seen as blockages between dominoes that prevent the fall of one affecting the next, thereby stopping the chain reaction. From the 1960s to the early 1980s, the barrier perspective gained new ground as a basis for accident prevention. Accidents were typically seen as a problem of uncontrolled transfer of harmful energy, and safety interventions were based on putting barriers between energy source and the object to be protected. The goal was to prevent, modify or mitigate the harmful effects of energy release, and pursuing it was instrumental in improving for example road safety. Strategies there ranged from reducing the amount of energy through speed limits, to controlling its release by salting roads or putting up side barriers, to absorbing energy with airbags (Rosness, Guttormsen, Steiro, Tinmannsvik, and Herrera, 2004).

The sequence-of-events model, and particularly its idea of accidents as the uncontrolled release and transfer of hazardous energy, connects effects to causes. For example in the

Columbia space shuttle accident, one can trace the energy effects from the energy of the foam strike, to the hole in the leading edge structure, to the heat build up during entry, and structural failure of the orbiter.

> ### CASE 3.1 SPACE SHUTTLE *COLUMBIA* BREAK-UP
>
> The physical causes of the loss of Space Shuttle *Columbia* in February 2003 can be meaningfully captured through a series of events that couples a foam strike not long after launch with the eventual breakup sequence during re-entry days later. A piece of insulating foam that had separated from the left bipod ramp section of the external tank at 81.7 seconds after launch struck the wing in the vicinity of the lower half of the reinforced carbon-carbon panel. This caused a breach in the Thermal Protection System on the leading edge of the left wing. During re-entry this breach in the Thermal Protection System allowed superheated air to penetrate through the leading edge insulation and progressively melt the aluminum structure of the left wing, resulting in a weakening of the structure until increasing aerodynamic forces caused loss of control, failure of the wing, and break-up of the Orbiter. This breakup occurred in a flight regime in which, given the current design of the Orbiter, there was no possibility for the crew to survive (*Columbia* Accident Investigation Report, 2003).

The result is a linear depiction of a causal sequence in terms of physical events and effects. The accident description is in terms of hazards specific to this physical system and physical environment (e.g., debris strikes and energy of entry). Such an analysis suggests ways to break the causal sequence by introducing or reinforcing defenses to prevent the propagation of the physical effects.

In sequence-of-events analyses people can assume a cause-consequence equivalence where each effect is also a cause, and each cause an effect, but also a symmetry between cause and effect. This has become an assumption that we often take for granted in our consideration of accidents. People may take for granted a symmetry between cause and effect, for example, that a very big effect (e.g., in numbers of fatalities) must have been due to a very big cause (e.g., egregious errors). The assumption of cause-consequence equivalence appears in discussions of accountability too, for example in how a judicial system typically assesses a person's liability or culpability on the basis of the gravity of the outcome (see Dekker, 2007).

But the sequence-of-events model is blind to patterns about cognitive systems and organizational dynamics. People only appear as another step that determined a branch or continuation of the sequence underway. Human performance becomes a discrete, binary event – the human did or did not do something – which failed to block the sequence or continued the sequence. These errors constitute a cause in the chain of causes/effects that led to the eventual outcome. Outsiders can easily construct alternative sequences, "the accident would have been avoided if only those people had seen or done this or that."

Versions of such thinking often show up in accident reports and remarks by stakeholders after accidents (e.g., How could the Mission Management Team have ignored the danger of the foam strike that occurred during launch? Why did NASA continue flying the Shuttle with a known problem?). This view of the human role leaves a vacuum, a vacuum that is filled by the suggestion careless people facilitated the physical sequence or negligently failed to stop the physical sequence. This dramatically oversimplifies the situations that people face at the time, often boiling things down to a choice between making an error or not making an error.

However, the *Columbia* Accident Investigation Board did not stop the analysis with a conclusion of human error. Instead, they "found the hole in the wing was produced not simply by debris, but by holes in organizational decision making" (Woods, 2003). The Board investigated the factors that produced the holes in NASA's decision making and found general patterns that have contributed to other failures and tragedies across other complex industrial settings (CAIB, 2003, Chapter 6; Woods, 2005):

○ Drift toward failure as defenses erode in the face of production pressure.
○ An organization that takes past success as a reason for confidence instead of investing in anticipating the changing potential for failure.
○ Fragmented distributed problem-solving process that clouds the big picture.
○ Failure to revise assessments as new evidence accumulates.
○ Breakdowns at the boundaries of organizational units that impede communication and coordination.

To deepen their analysis the *Columbia* Board had to escape hindsight bias and examine the organizational context and dynamics such as production pressure that led management to see foam strikes as a turn around issue and not as a safety of flight issue.

THE RELATIONSHIP BETWEEN INCIDENTS AND ACCIDENTS

An important by-product of the sequence-of-events model is the relationship between incidents and accidents. If an accident is the conclusion of a sequence of events that proceeded all the way to failure, then an incident is a similar progression with one difference – it was stopped in time. This has been an attractive proposition for work on safety – arrest a causal progression early on by studying more frequently occurring incidents to identify what blocks or facilitates the accident sequence. The assumption is that incidents and accidents are similar in substance, but only different in outcome: the same factors contribute to the progression towards failure, but in one case the progression is stopped, in the other it is not. One example is after an accident occurs investigators notice that there had been previous "dress rehearsals" where the same problem had occurred but had been recovered from before negative consequences.

Take air traffic control, for example. An accident would be a mid-air collision with another aircraft. An incident would be the violation of separation minima (e.g., 5 nautical miles lateral and 1,000 feet vertical) but no physical contact between aircraft. A near miss would be coming close to violating the minimum separation criterion. One can then look for actions or omissions that appear to increase the risk of a near miss or incident – so-called unsafe acts.

The iceberg model assumes that these categories are directly related in frequency and causality. Unsafe acts lead to near misses, near misses to incidents, incidents to accidents in the same causal progression. The iceberg model proposes that there are a certain number of incidents for each accident, and a certain number of near misses for each incident, and so forth. The typical ratio used is 1 accident for 10 incidents for 30 near misses for 600 unsafe acts (1:10:30:600).

As systems become more complex and as their operations become safer (e.g., air traffic control, nuclear power generation, commercial aviation), the assumptions behind the iceberg model become increasingly questionable and the relationships between incidents and accidents more complex (Amalberti, 2001; Hollnagel, 2004).

Data from scheduled airline flying in the US illustrate the difficulties. Table 3.1 from Barnett and Wang (2000) shows correlations between the number of nonfatal accidents or incidents per 100,000 major carrier departures and their passenger mortality risk. Interestingly, all correlations are negative: carriers with *higher* rates of non-fatal accidents or non-fatal incidents had *lower* passenger mortality risks. This directly contradicts the iceberg proposition: the more incidents there are, the fewer fatal accidents. In fact, the table basically inverts the iceberg, because correlations become increasingly negative when the events suffered by the carrier become more severe. If the non-fatal accident that happened to the carrier is more severe, in other words, there is even less chance that a passenger will die onboard that carrier.

Table 3.1 Correlations between the number of nonfatal accidents or incidents per 100,000 major US jet air carrier departures and their passenger mortality risk (January 1, 1990 to March 31, 1996 (Barnett and Wang, 2000, p. 3))

Type of non-fatal event	Correlation
Incidents only	-0.10
Incidents and accidents	-0.21
Accidents only	-0.29
Serious accidents only	-0.34

Statistically, scheduled airline flying in the US is very safe (a single passenger would have to fly 19,000 years before the expected probability of a death in an airline accident). Amalberti (2001) notes the paradox of ultra-safe systems: it is particularly in those systems with very low overall accident risk that the predictive value of incidents becomes very small. In such ultra-safe systems:

> Accidents are different in nature from those occurring in safe systems: in this case accidents usually occur in the absence of any serious breakdown or even of any serious error. They result from a combination of factors, none of which can alone cause an accident, or even a serious incident; therefore these combinations remain difficult to

detect and to recover using traditional safety analysis logic. For the same reason, reporting becomes less relevant in predicting major disasters. (Amalberti, 2001, p. 112)

Detailed investigations of accidents thus frequently show that the system was managed *towards* catastrophe, often for a long while. Accidents are not anomalies that arise from isolated human error. Instead, accidents are "normal" events that arise from deeply embedded features of the systems of work (Perrow, 1984). Complex systems have a tendency to move incrementally towards the boundaries of safe operations (Rasmussen, 1997; Cook and Rasmussen, 2005). Because they are expensive to operate, there is a constant drive to make their operations cheaper or more efficient. Because they are complex, it is difficult to project how changes in the operations will create opportunities for new forms of failure.

CASE 3.2 TEXAS A&M UNIVERSITY BONFIRE COLLAPSE

The Texas A&M University bonfire tragedy is a case in point. The accident revealed a system that was profoundly out of control and that had, over a long period, marched towards disaster (see Petroski, 2000; Linbeck, 2000). On November 18, 1999, a multi-story stack of logs that was to be burned in a traditional football bonfire collapsed while being built by students at the Texas A&M University. Twelve students working on the structure were crushed to death as the structure collapsed. Twenty-seven others were injured. The casualties overwhelmed the medical facilities of the area. It was the worst such disaster at a college campus in the United States and was devastating within the tight knit university community that prided itself on its engineering college.

The bonfire was a Texas A&M football tradition that extended over many years. It began in 1928 as a haphazard collection of wooden palettes. It grew gradually, increasing in scale and complexity each year until the 1990s when it required a crane to erect. In 1994 a partial collapse occurred but was attributed to shifting ground underneath the structure rather than structural failure per se (in addition, actions were taken to control drinking by students participating in the project).

An independent commission was established to investigate the causes of the collapse (Linbeck, 2000). Extensive and expensive engineering studies were conducted that showed that the collapse was the result of specific aspects of the design of the bonfire structure (one could focus on the physical factors that led to the collapse, for example wedging, internal stresses, bindings). The investigation revealed that the collapse happened because the bonfire had evolved into a large scale construction project over a number of years but was still built largely by unsupervised amateurs. The accident was an organizational failure; as the structure to be built had evolved in scale and complexity, the construction went on without proper

> design analyses, engineering controls, or proactive safety analyses. The accident report concluding statement was:
>
> *"Though its individual components are complex, the central message is clear. The collapse was about physical failures driven by organizational failures, the origins of which span decades of administrations, faculty, and students. No single factor caused the collapse, just as no single change will ensure that a tragedy like this never happens again."*

Accidents such as the Texas A&M bonfire collapse illustrate the limits of the sequence-of-events model. The relative lack of failure over many years produced a sense that failure was unlikely. The build up of structural and engineering complexity was gradual and incremental. No group was able to recognize the changing risks and introduce the engineering controls and safety processes commensurate with the scale of construction (or accept the costs in time, human resources, and money these changes required). Precursor events occurred and were responded to, but only with respect to the specific factors involved in that event. The partial collapse did not trigger recognition that the scale change required new engineering, control, and safety processes. The bonfire structures continued to grow in scale and complexity until the structure of the late 1990s reached new heights without anyone understanding what those heights meant. Only after the disaster did the risks of the situation become clear to all stakeholders.

In this case, as in others, there was a slow steady reduction of safe operating margins over time. Work proceeded under normal everyday pressures, expectations and resources. A record of past success obscured evidence of changing risks and provided an unjustified basis for confidence in future results.

MAN-MADE DISASTER THEORY

In 1978, Barry Turner offered one of the first accounts of accidents as a result of normal, everyday organizational decision making. Accidents, Turner concluded, are neither chance events, nor acts of God, nor triggered by a few events and unsafe human acts. Nor is it useful to describe accidents in terms of the technology in itself (Pidgeon and O'Leary, 2000). Turner's idea was that "man-made disasters" often start small, with seemingly insignificant operational and managerial decisions. From then, there is an *incubation period*. Over a long time, problems accumulate and the organization's view of itself and how it manages its risk grows increasingly at odds with the actual state of affairs (miscalibration), until this mismatch actually explodes into the open in the form of an accident (Turner, 1978). Man-made disaster theory preserved important notions of the sequence-of-events model (e.g., problems at the root that served to trigger others over time) even if the sequence spread further into the organization and deeper into history than in any previous model of accidents. Yet the Turner's insights added a new focus and language to the arsenal of safety thinking.

An important post-accident discovery highlighted by man-made disaster theory is that seemingly innocuous organizational decisions turned out to interact, over time, with other preconditions in complex and unintended ways. None of those contributors alone is likely to trigger the revelatory accident, but the way they interact and add up falls outside the predictive scope of people's model of their organization and its hazard control up to that moment. Turner's account was innovative because he did not define accidents in terms of their physical impact (e.g., uncontrolled energy release) or as a linear sequence of events. Rather, he saw accidents as organizational and sociological phenomena. Accidents represent a disruption in how people believe their system operates; a collapse of their own norms about hazards and how to manage them. An accident, in other words, comes as a shock to the image that the organization has of itself, of its risks and of how to contain them. The developing vulnerability is hidden by the organization's belief that it has risk under control.

Stech (1979) applied this idea to the failure of Israeli intelligence organizations to foresee the Yom Kippur war, even though all necessary data that pointed in that direction was available somewhere across the intelligence apparatus. Reflecting on the same events, Lanir (1986) used the term "fundamental surprise," to capture this sudden revelation that one's perception of the world is entirely incompatible with reality. Part V, particularly Chapter 14, of this book details the fundamental surprise process as part of regularities about how organizations learn and fail to learn from accidents and before accidents. Interestingly, the surprise in man-made disaster theory is not that a system that is normally successful suddenly suffers a catastrophic breakdown. Rather, the surprise is that a successful system produces failure as a systematic by-product of how it normally works.

Take the processing of food in one location. This offers greater product control and reliability (and thus, according to current ideas about food safety, better and more stringent inspection and uniform hygiene standards). Such centralized processing, however, also allows effective, fast, and wide-spread distribution of unknowingly contaminated food to many people at the same time precisely thanks to the existing systems of production. This happened during the outbreaks of food poisoning from the *E-coli* bacterium in Scotland during the 1990s (Pidgeon and O'Leary, 2000). This same centralization of food preparation, now regulated and enforced as being the safest in many Western military forces, also erodes cooks' expertise at procuring and preparing local foods in the field when on missions outside the distribution reach of centrally prepared meals. As a result of centralized control over food preparation and safety, the incidence of food poisoning in soldiers on such missions typically has gone up.

CASE 3.3   BOEING 757 LANDING INCIDENT

On the evening of December 24, 1997, the crew of a Boeing 757 executed an autopilot-coupled approach to a southerly runway at Amsterdam (meaning the autopilot was flying the aircraft down the electronic approach path toward the runway). The wind was very strong and gusty out of the south-west. The pilot disconnected the autopilot at approximately 100 ft above the ground in order to make a manual landing. The aircraft touched down hard with its right main wheels first. When the nose gear touched down hard with the aircraft in a crab angle (where the airplane's body is moving slightly sideways along the runway so as to compensate for the strong crosswind), the nose wheel collapsed, which resulted in serious damage to the electric/electronic systems and several flight- and engine control cables. The aircraft slid down the runway, was pushed off to the right by the crosswind and came to rest in the grass next to the runway. All passengers and crew were evacuated safely and a small fire at the collapsed nose wheel was quickly put out by the airport fire brigade (Dutch Safety Board, 1999).

In an effort to reduce the risks associated with hard landings, runway overruns and other approach and landing accidents, the aviation industry has long championed the idea of a "stabilized approach" whereby no more large changes of direction, descent rate, power setting and so forth should be necessary below a particular height, usually 500 feet above the ground. Should such changes become necessary, a go-around is called for. Seeing a stabilized approach as "the accepted airmanship standard" (Sampson, 2000), many airline safety departments have taken to stringent monitoring of electronic flight data to see where crews may not have been stabilized (yet "failed" to make a go-around). One operational result (even promoted in various airlines' company procedures) is that pilots can become reluctant to fly approaches manually before they have cleared the 500 ft height window: rather they let the automation fly the aircraft down to at least that height. This leads to one of the ironies of automation (Bainbridge, 1987): an erosion of critical manual control skills that are still called for in case automation is unable to do the job (as it would have been unable to land the aircraft in a crosswind as strong as at Amsterdam at that time). Indeed, the investigation concluded how there would have been very little opportunity from 100 ft on down for the pilot to gain effective manual control over the aircraft in that situation (Dutch Safety Board, 1999).

Some would call this "pilot error." In fact, one study cited in the investigation concluded that most crosswind-related accidents are caused by improper or incorrect aircraft control or handling by pilots (van Es, van der Geest and Nieuwpoort, 2001). But Man-made disaster theory would say the pilot error was actually a non-random effect of a system of production that helped the industry achieve success according to its dominant model of risk: make sure approaches are stabilized, because non-stabilized approaches are a major source of hazard. This seemingly sensible and safe strategy incubated an unintended vulnerability as a side effect. Consistent with Turner's ideas, the accident revealed the limits of the model of success and risk used by the organization (and by extension, the industry).

MAN-MADE DISASTER THEORY AND "HUMAN ERROR"

Man-made disaster theory holds that accidents are administrative and managerial in origin – not just technological. A field that had been dominated by languages of energy transfers and barriers was thus re-invigorated with a new perspective that extended genesis of accidents into organizational dynamics that developed over longer time periods. "Human errors" that occurred as part of proximal events leading up to the accident are the result of problems that have been brewing inside the organization over time. Since Turner accident inquiries have had to take organizational incubation into account and consider the wider context from which "errors" stem (often termed "latent failures"). The theory was the first to put issues of company culture and institutional design (which determines organizational cognition, which in turn governs information exchange) at the heart of the safety question (Pidgeon and O'Leary, 2000). Basically all organizational accident models developed since the seventies owe intellectual debt to Turner and his contemporaries (Reason, Hollnagel and Pariès, 2006), even if the processes by which such incubation occurs are still poorly understood (Dekker, 2005).

There is an unresolved position on human error in Man-made disaster theory and the subsequent models it has inspired. The theory posits that "despite the best intentions of all involved, the objective of safely operating technological systems could be subverted by some very familiar and 'normal' processes of organizational life" (Pidgeon and O'Leary, 2000, p. 16). Such "subversion" occurs through usual organizational phenomena such as information not being fully appreciated, information not correctly assembled, or information conflicting with prior understandings of risk. Turner noted that people were prone to discount, neglect or not take into discussion relevant information, even when available, if it mismatched prior information, rules or values of the organization. Thus, entire organizations could fail to take action on danger signals because of what he called "decoy" phenomena that distracted from the building hazard (Rosness et al., 2004).

The problem is that it doesn't explain how people in management do not fully appreciate available information despite good intentions of all involved. There is a need to explain why some interpretations seemed right at the time, despite other information (but see Part III on Operating at the Sharp End). Neither Man-made disaster theory, nor do its offshoots (e.g., Reason, 1997), offer a solution. "Not fully" appreciating information implies a norm of what "fully" would have been. Not "correctly" assembling information implies a norm of what "correct" assembly would be. These norms, however, are left unexpressed in the theory because they exist only in hindsight, from the point of view of an omniscient retrospective observer.

In other work on how organizations deal with information, Westrum (1993) identified three types of organizational culture that shapes the way people respond to evidence of problems:

○ **Pathological culture**. Suppresses warnings and minority opinions, responsibility is avoided and new ideas actively discouraged. Bearers of bad news are "shot," failures are punished or covered up.
○ **Bureaucratic culture**. Information is acknowledged but not dealt with. Responsibility is compartmentalized. Messengers are typically ignored because new ideas are seen as problematic. People are not encouraged to participate in improvement efforts.
○ **Generative culture**. Is able to make use of information, observations or ideas wherever they exist in the system, without regard to the location or status of the

person or group having such information, observation or ideas. Whistleblowers and other messengers are trained, encouraged and rewarded.

Westrum's generative culture points to some of the activities that can make teams and organizations resilient – able to remain sensitive to the possibility of failure and constantly updating their models of risk so they can adapt effectively under pressure, even in the face of novelty.

The other problem in Man-made disaster theory is that it just shifts the referent for human error to other people in other roles at the blunt end of the system. It relocates the problem of human error further up an extended causal pathway that includes roles and factors away from the place and time of the accident itself. It shifts the explanation for the accident from the sharp-end operators (who inherit the accident rather than cause it) but places it on other people (managers who failed to appreciate information about risks) earlier on).

This is a serious difficulty to be avoided. Expanding the analysis of contributing factors appears to explain one human error (operator error) by referring to another (manager error) and then stopping there. The same difficulty arises in Part IV where we examine how clumsy technology can induce erroneous actions and assessments. The human-technology interaction factors can be mis-interpreted as shifting the issue from operator error to designer error. Shifting from sharp end human error to blunt end human error is a failure of systems thinking.

THE LATENT FAILURE MODEL (AKA "SWISS CHEESE")

The latent failure model is an evolution and combination of ideas from preceding theories and models on accident causation, particularly the sequence-of-events model and Man-made disasters theory. According to the latent failure model, which first appeared in developed form in Reason (1990), disasters are characterized by a concatenation of several small failures and contributing events – rather than a single large failure. Multiple contributors are all necessary but individually insufficient for the disaster to occur. For example, the combination of multiple contributing events is seen in virtually all of the significant nuclear power plant incidents, including Three Mile Island, Chernobyl, the Brown's Ferry fire, the incidents examined in Pew et al. (1981), the steam generator tube rupture at the Ginna station (Woods, 1982) and others. In the near miss at the Davis-Besse nuclear station (NUREG-1154), about 10 machine failures and several erroneous human actions were identified that initiated the loss-of-feedwater accident and determined how it evolved.

Some of the factors that combine to produce a disaster are latent in the sense that they were present before the incident began. Turner (1978) discussed this in terms of the incubation of factors prior to the incident itself, and Reason (1990) refers to hidden pathogens that build in a system in an explicit analogy to viral processes in medicine. Reason (1990) uses the term latent failure to refer to errors or failures in a system that produce a negative effect but whose consequences are not revealed or activated until some other enabling condition is met. A typical example is a failure that makes safety systems unable to function properly if called on, such as the maintenance failure that resulted in the emergency feedwater system being unavailable during the Three Mile Island incident (The Kemeny Commission, 1979). Latent failures require a trigger, that is, an initiating or enabling event, that activates its effects or consequences. For example in the space shuttle *Challenger* disaster, the decision to launch in cold weather was the initiating event that activated the consequences of the latent failure in booster seal design (Rogers et al., 1986).

The concatenation of factors in past disasters includes both human and machine elements intertwined as part of the multiple factors that contribute to incident evolution. One cannot study these as separate independent elements, but only as part of the dynamics of a human-machine operational system that has adapted to the demands of the field of activity and to the resources and constraints provided by the larger organizational context (Rasmussen, 1986). The latent failure model thus distinguishes between active and latent failures:

○ Active failures are "unsafe acts" whose negative consequences are immediately or almost immediately apparent. These are associated with the people at the "sharp end," that is, the operational personnel who directly see and influence the process in question.
○ Latent failures are decisions or other issues whose adverse consequences may lie dormant within the system for a long time, only becoming evident when they combine with other factors to breach the system's defenses (Reason, 1990). Some of the factors that serve as "triggers" may be active failures, technical faults, or atypical system states. Latent failures are associated with managers, designers, maintainers, or regulators – people who are generally far removed in time and space from handling incidents and accidents.

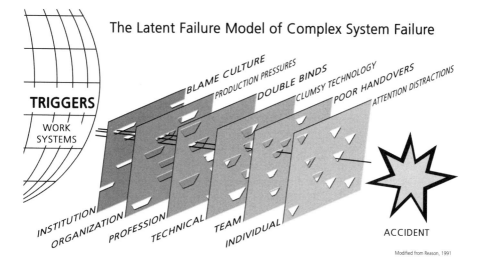

Figure 3.1 Complex systems failure according to the latent failure model. Failures in these systems require the combination of multiple factors. The system is defended against failure but these defenses have defects or "holes" that allow accidents to occur

CASE 3.4   THE AIR ONTARIO FLIGHT 1363 ACCIDENT AT DRYDEN, CANADA

On Friday, March 10, 1989, a Fokker F-28 commuter jet aircraft took off from Dryden, Ontario on the last leg of a series of round trips between Winnepeg, Manitoba and Thunder Bay, Ontario. During the brief stopover in Dryden, the aircraft had been refueled. The temperature was hovering near freezing and it had been raining or snowing since the aircraft had landed. Several passengers and at least one crew member had noticed that slush had begun to build up on the wings. Flight 1363 began its takeoff roll but gathered speed slowly and only barely cleared the trees at the end of the runway. The Fokker never became fully airborne but instead crashed less than 1 km beyond the end of the runway. The aircraft, loaded with fuel, was destroyed by fire. Twenty-four people on board, including the pilot and co-pilot, were killed.

The initial assessment was that pilot error, specifically the decision to take off despite the icy slush forming on the wings, was the cause of the accident. The inexplicable decision to attempt takeoff was not in keeping with the pilot's record or reputation. The pilot was experienced and regarded by others as a thoughtful, cautious, and competent man who operated "by-the-book." Nevertheless, it was immediately obvious that he had chosen to take off in the presence of hazardous conditions that a competent pilot should have known were unacceptable. The reactions to this particular accident might well have ended there. Human error by practitioners is well known to be the proximate cause of 70% of accidents. At first, the Dryden Air Ontario crash seemed to be just another instance of the unreliability of humans in technological settings. If the pilot had been inexperienced or physically or mentally impaired (e.g., drinking alcohol before the crash), it is likely that attention would have turned away and the crash would today be remembered only by the survivors and families of those who died.

Instead, the Canadian Federal government commissioned an unprecedented investigation, under the direction of retired Supreme Court Justice Moshansky into all the factors surrounding the accident. Why it did so is not entirely clear but at least three factors seem to have played roles. First, the scale of the event was too large to be treated in ordinary ways. Images of the charred wreckage and the first-hand accounts of survivors captivated the entire country. The catastrophe was national in scale. Second, the accident "fit" hand-in-glove into a set of concerns about the state of Canadian aviation, concerns that had been growing slowly over several years. These years had seen substantial changes in Canadian commercial aviation due to airline deregulation. New aircraft were being brought into service and new routes were being opened. In addition, the aviation industry itself was in turmoil. Once-small companies were expanding rapidly and larger companies were buying up smaller ones. Significantly, instead of keeping pace with the growth of commercial aviation, the Canadian government's aviation regulatory oversight body was shrinking as the government sought to reduce its budget deficit. There was no obvious connection between any of these large scale factors and the accident at Dryden, Ontario.

The investigation took almost two years to complete and became the most exhaustive, most extensive examination of an aviation accident ever conducted (Moshansky, 1992). Well over 200,000 pages of documents and transcripts were collected and analyzed. The investigators explored not just the mechanics of flight under icing conditions but details about the running of the airport, the air transportation system, its organization, and its regulation. All these factors were linked together in the Commission's four-volume report. The Report does not identify a single cause or even multiple causes of the accident. Instead, it makes clear that the aviation system contained many faults that together created an environment that would eventually produce an accident – if not on the 10th day of March in Dryden, Ontario, then on some other day in some other place (Maurino et al., 1999).

Bad weather at the Dryden airport was just one of many problems that came together on March 10, 1989. The airline itself was a family operation without strong management. It had traditionally relied on smaller, prop aircraft and had only recently begun jet operations. The operating manual for the Fokker F-28 had not yet been approved by Canadian regulators. The company's safety manager, an experienced pilot, had recently resigned because of disputes with management. There were 'deferred' maintenance items, among them fire sensors in the small engine the Fokker carried that would allow it to start its main engines. Company procedures called for the engines to be shut down for deicing of the wings but there was no convenient way to restart them at Dryden: the company did not have ground starting equipment for its new jet aircraft at the Dryden airport. To deice the aircraft would have required turning off the engines but once they were turned off there was no way to restart them (The F-28 aircraft had an auxiliary starting engine located in the tail to allow the aircraft to start its own jet engines using internal power. This engine was believed by the pilots to be unusable because certain sensors were not working. In fact, the auxiliary engine was operable.) Bad weather at Dryden caused snow and ice to build up on the aircraft wings. The Dryden refueling was necessary because the airline management had required the pilots to remove fuel before taking off on from Thunder Bay, Ontario for the trip to Winnipeg. The pilot had wanted to leave passengers behind in Thunder Bay to avoid the need to refuel but management had ordered him to remove fuel instead, creating the need for refueling in Dryden (the situation was even more complex than indicated here and involves weather at expected and alternate airports, the certification of the pilot for operation of the Fokker, and detailed characteristics of the aircraft. For a complete description, see Moshansky, 1992). The takeoff of Flight 1363 from Dryden was further delayed when a single engine airplane's urgently requested use of the one runway in order to land because the snow was making visibility worse. Ultimately, over 30 contributing factors were identified, including characteristics of the deregulation of commercial aviation in Canada, management deficiencies in the airline company, and lack of maintenance and operational equipment.

None of these problems by itself was sufficient in itself to cause a crash. Only in combination could these multiple latent conditions create the conditions needed for the crash. In hindsight, there were plenty of opportunities to prevent the accident. But the fact that the multiple flaws are necessary to create the disaster has the paradoxical effect of making each individual flaw seem insignificant. Seen in isolation, no one flaw appears dangerous. As a result, many such flaws may accumulate within a system without raising alarm. When they combine, they present the operators with a situation that teeters on the very edge of catastrophe. This was the situation in the case of Flight 1363.

The pilots were not so much the instigators of the accident as the recipients of it. Circumstances had combined (some would say conspired) to create a situation that was rife with pressures, uncertainty, and risk. The pilots were invited to manage their way out of the situation but were offered no attractive opportunities to do so. Rather than being a choice between several good alternatives, the system produced a situation where the pilots were forced to choose between bad alternatives under conditions of uncertainty. They made an effort to craft a safe solution but were obstructed by managers who insisted that they achieve production goals.

Unlike many post-accident inquiries, the investigation of the crash of Flight 1393 was detailed and broad enough to show how the situation confronting the pilots had arisen. It provided a fine-grain picture of the kinds of pressures and difficulties that operators at the sharp end of practice confront in daily work. It showed how the decisions and actions throughout the aviation system had brought these pressures and difficulties together in the moments before the crash. This is now recognized as a "systems view" of the accident. It is a picture of the system that shows, in detail, how the technical characteristics of the workplace, the technical work that takes place there, and the pressures and difficulties that the workers experience combine to create the situation that produced the accident. Rather than attributing the accident to a discrete cause or causes, the investigation works towards providing a detailed account of the interactions between the factors that created the situation. In itself, the latent failure model could not capture all this, but it has been a very important contribution to making many more stakeholders think more critically about the rich context that surrounds and helps produce accidents.

THE LATENT FAILURE MODEL AND "HUMAN ERROR"

The latent failure model has helped redirect the focus away from front-line operators and towards the upstream conditions that influenced and constrained their work. As Reason put it in 1990:

> Rather than being the main instigators of an accident, operators tend to be the inheritors of system defects created by poor design, incorrect installation, faulty maintenance and bad management decisions. Their part is usually that of adding the final garnish to a lethal brew whose ingredients have already been long in the cooking. (p. 173)

Several chapters in this book shows how blunt end factors can shape practitioner cognition and create the potential for erroneous actions and assessments. They will also show how the clumsy use of technology can be construed as one type of latent failure. This type of latent failure arises in the design organization. It predictably leads to certain kinds of unsafe acts on the part of practitioners at the sharp end and contributes to the evolution of incidents towards disaster. Task and environmental conditions are typically thought of as "performance shaping factors." The latent failure model, then, has provided an orderly set of concepts for accident analysts and others to consider when they want to find out what lies behind the label "human error."

According to the latent failure model, we should think of accident potential in terms of organizational processes, task and environmental conditions, individual unsafe acts, and failed defenses (see Figure 3.1). A safety-critical system is surrounded by defenses-in-depth (as depicted by the various layers between hazards and the object or process to be protected in the figure). Defenses are measures or mechanisms that protect against hazards or lessen the consequences of malfunctions or erroneous actions. Some examples include safety systems or forcing functions such as interlocks. According to Reason (1990), the "best chance of minimizing accidents is by identifying and correcting these delayed action failures (latent failures) before they combine with local triggers to breach or circumvent the system's defenses." This is consistent with original 1960s ideas about barriers and the containment of unwanted energy release (Hollnagel, 2004; Rosness et al., 2004).

None of these layers are perfect, however, and the "holes" in them represent those imperfections. The organizational layer, for example, involves such processes as goal setting, organizing, communicating, managing, designing, building, operating, and maintaining. All of these processes are fallible, and produce the latent failures that reside in the system. This is not normally a problem, but when combined with other factors, they can contribute to an accident sequence. Indeed, according to the latent failure model, accidents happen when all of the layers are penetrated (when all their imperfections or "holes" line up). Incidents, in contrast, happen when the accident progression is stopped by a layer of defense somewhere along the way. This idea is a carry over from the earlier sequence-of-events model, as is the linear depiction of a failure progression.

The latent failure model broadens the story of error. It is not enough to stop with the attribution that some individual at the sharp end erred. The concept of latent failures

CASE 3.5 EASTERN AIRLINES L1011 FROM MIAMI TO NASSAU, MAY 1983

The aircraft lost oil pressure in all three of its engines in mid-flight. Two of the engines stopped, and the third gave out at about the time the crew safely landed the aircraft. The proximal event was that O-rings, which normally should be attached to an engine part, were missing from all three engines (It is interesting to note that from the perspective of the pilot, it seemed impossible that all three should go out at once. There must have been a common mode failure – but what was it? The only thing they could think of was that it must be an electrical system problem. In actuality, it was a common mode failure, though a different one than they hypothesized). A synopsis of relevant events leading up to the incident is given below, based on the National Transportation Safety Board report (NTSB, 1984) and on Norman's commentary on this incident (Norman, 1992).

One of the tasks of mechanics is to replace an engine part, called a master chip detector, at scheduled intervals. The master chip detector fits into the engine and is used to detect engine wear. O-rings are used to prevent oil leakage when the part is inserted. The two mechanics for the flight in question had always gotten replacement master chip detectors from their foreman's cabinet. These chip detectors were all ready to go, with new O-rings installed. The mechanics' work cards specified that new O-rings should be installed with a space next to this instruction for their initials when the task was completed. However, their usual work situation meant that this step was unnecessary, because someone else (apparently their supervisor) was already installing new O-rings on the chip detectors.

The night before the incident, an unusual event occurred. When the mechanics were ready to replace master chip detectors, they found there were no chip detectors in the foreman's cabinet. The mechanics had to get the parts from the stockroom. The chip detectors were wrapped in a "semi-transparent sealed plastic package with a serviceable parts tag." The mechanics took the packages to the aircraft and replaced the detectors in low light conditions. It turned out the chip detectors did not have O-rings attached. The mechanics had not checked for them, before installing them. There was a check procedure against improper seals: motoring the engines to see if oil leaked. The technicians did this, but apparently not for a long enough time to detect oil leaks.

One might argue that the technicians should have checked the O-rings on the part, especially since they initialed this item on the work card. But consider that they did not strictly work from the work card – the work card said that they should install a new seal. But they never needed to; someone else always took care of this, so they simply checked off on it. Also, they could not work strictly from procedure; for example, the work card read "motor engine and check chip detector for leaks" but it didn't specify how long. The mechanics had to fill in the gap, and it turned out the time they routinely used was too short to detect leaks (a breakdown in the system for error detection).

Even without these particular technicians, the system held the potential for breakdown. Several problems or latent failures existed. The unusual event (having to get the part from supply) served as a trigger. (These latent failures are points where a difference might have prevented this particular incident.) Some of these were:

(a) The fact that someone other than the technicians normally put the O-rings on the chip detectors left in the cabinet and yet did not initial the work card, effectively leaving no one in charge of O-ring verification. (There would have been no place to initial since the task of using a new seal was a subtask of the larger step which included replacing the chip detector.)
(b) The fact that the chip detectors from supply were not packed with O-rings.
(c) Personnel did not know what was a sufficient length of time to run the engines to see if their tasks had been carried out successfully.

Other factors that may have played a role include:

(a) Low lighting conditions and the necessity of working by feel when inserting the part made it unlikely that the lack of O-rings would have been detected without explicitly checking for them.
(b) Special training procedures concerning the importance of checking O-rings on the chip detectors were posted on bulletin boards and kept in a binder on the general foreman's desk. Theoretically, the foremen were supposed to ensure that their workers followed the guidance, but there was no follow-up to ensure that each mechanic had read these.
(c) The variation from a routine way of doing something (opening up the potential for slips of action).

The latent factors involved multiple people in different jobs and the procedures and conditions established for the tasks at the sharp end. Notice how easy it is to miss or rationalize the role of latent factors in the absence of outcome data (see Part V for more on this point). In this case, the airline had previous O-ring problems, but these were attributed to the mechanics. According to the NTSB report, the propulsion engineering director of the airline, after conferring with his counterparts, said that all the airlines were essentially using the same maintenance procedure but were not experiencing the same in-flight shutdown problems. Hence, it was concluded that the procedures used were valid, and that the

> problems in installation were due to personnel errors. Also, in reference to the eight incidents that occurred in which O-rings were defective or master chip detectors were improperly installed (prior to this case), the "FAA concluded that the individual mechanic and not Eastern Air Lines maintenance procedures was at fault" (National Transportation Safety Board, 1984).
>
> As Norman (1992) points out, these are problems in the system. These latent failures are not easy to spot; one needs a systems view (i.e., view of the different levels and their interactions) as well as knowledge of how they hold the potential for error. Because of how difficult it is to see these, and how much easier it is to focus on the individual and the actions or omissions that directly impacted the event, the tendency is to attribute the problem to the person at the sharp end. But behind the label "human error" is another story that points to many system-oriented deficiencies that made it possible for the faulty installation to occur and to go undetected.

highlights the importance of organizational factors. It shows how practitioners at the sharp end can be constrained or trapped by larger factors. Even though it throws the net much wider, encompassing a larger number than factors than may have been usual, the latent failure model holds on to a broken-component explanation of accidents. The latent failures themselves are, or contribute to, (partially) broken layers of defense, for example. The model proposes that latent failures can include organizational deficiencies, inadequate communications, poor planning and scheduling, inadequate control and monitoring, design failures, unsuitable materials, poor procedures (both in operations and maintenance), deficient training, and inadequate maintenance management (Reason, 1993). The problem is that these are all different labels for "human error," even if they refer to other kinds of errors by other people inside or outside the organization. Explaining operator error by referring to errors by other people fails to adopt systems thinking (Rosness et al., 2005).

The latent failure model also reserves a special place for violations. These, according to the model, are deviations from some code of practice or procedure. Stakeholders often hugely overestimate the role of such "violations" in their understanding of accidents ("if only operators followed the rules, then this would never have happened") and can presume that local adaptations to rules or other written guidance were unique to that situation, the people in it or the outcome it produced. This is not often the case: written guidance is always underspecified relative to the actual work-to-be-performed, as well as insensitive to many changes in context, so people always need to bridge the gaps by interpreting and adapting. To understand failure and success in safety-critical worlds where multiple goals compete for people's attention and resources are always limited, it may not be helpful to see adaptations in such a strong normative light, where the rule is presumed right and the operator always wrong. In the sections on practitioner tailoring and on rule following in Part III, we take a systems view of procedures, brittleness, and adaptation.

The best chance of minimizing accidents is by learning how to detect and appreciate the significance of latent failures before they combine with other contributors to produce disaster (Reason, 1990). But this is where the depiction of a complex system as a static set of layers presents problems. It does not explain how such latent failures come into being, nor how they actually combine with active failures. Also, the model does not tell how layers of defense are gradually eroded, for example under the pressures of production and resource limitations and over-confidence based on successful past outcomes.

4

COMPLEXITY, CONTROL AND SOCIOLOGICAL MODELS

NORMAL ACCIDENT THEORY

Highly technological systems such as aviation, air traffic control, telecommunications, nuclear power, space missions, and medicine include potentially disastrous failure modes. These systems, consistent with the barrier idea in the previous chapter, usually have multiple redundant mechanisms, safety systems, and elaborate policies and procedures to keep them from failing in ways that produce bad outcomes. The results of combined operational and engineering measures make these systems relatively safe from single point failures; that is, they are protected against the failure of a single component or procedure directly leading to a bad outcome. But the paradox, says Perrow (1984), is that such barriers and redundancy can actually add complexity and increase opacity so that, when even small things start going wrong, it becomes exceptionally difficult to get off an accelerating pathway to system breakdown. The need to make these systems reliable, in other words, also makes them very complex. They are large systems, semantically complex (it generally takes a great deal of time to master the relevant domain knowledge), with tight couplings between various parts, and operations are often carried out under time pressure or other resource constraints.

Perrow (1984) promoted the idea of *system accidents*. Rather than being the result of a few or a number of component failures, accidents involve the unanticipated interaction of a multitude of events in a complex system – events and interactions whose combinatorial explosion can quickly outwit people's best efforts at predicting and mitigating disaster. The scale and coupling of these systems creates a different pattern for disaster where incidents develop or evolve through a conjunction of several small failures. Yet to Normal Accidents Theory, analytically speaking, such accidents need not be surprising at all (not even in a fundamental sense). The central thesis of what has become known as normal accident theory (Perrow, 1984) is that accidents are the structural and virtually inevitable product of systems that are both interactively complex and tightly coupled. Interactive complexity and coupling are two presumably different dimensions along which Perrow plotted a number of systems (from manufacturing to military operations to nuclear power plants). This separation into two dimensions has spawned a lot of thinking and discussion (including whether they are separable at all), and has offered new ways of looking at how to manage and control complex, dynamic technologies, as well as suggesting what may lie behind the label "human error" if things go wrong in a tightly coupled, interactively complex system. Normal accident theory predicts that the more tightly coupled and complex a system is, the more prone it is to suffering a "normal" accident.

Interactive complexity refers to component interactions that are non-linear, unfamiliar, unexpected or unplanned, and either not visible or not immediately comprehensible for people running the system. Linear interactions are those in expected and familiar production or maintenance sequences, and those that are quite visible and understandable even if unplanned. Complex interactions are those of unfamiliar sequences, or unplanned and unexpected sequences, and either not visible or not immediately comprehensible (Perrow, 1984). An electrical power grid is an example of an interactively complex system. Failures, when they do occur, can cascade through these systems in ways that may confound the people managing them, making it difficult to stop the progression of failure (this would also go for the phone company AT&T's Thomas Street outage, even if stakeholders implicated "human error").

In addition to being either linearly or complexly interactive, systems can be loosely or tightly coupled. They are tightly coupled if they have more time-dependent processes (meaning they can't wait or stand by until attended to), sequences that are invariant (the order of the process cannot be changed) and little slack (e.g., things cannot be done twice to get it right). Dams, for instance, are rather linear systems, but very tightly coupled. Rail transport is too. In contrast, an example of a system that is interactively complex but not very tightly coupled is a university education. It is interactively complex because of specialization, limited understanding, number of control parameters and so forth. But the coupling is not very tight. Delays or temporary halts in education are possible, different courses can often be substituted for one another (as can a choice of instructors), and there are many ways to achieving the goal of getting a degree.

CASE 4.1   A COFFEE MAKER ONBOARD A DC-8 AIRLINER

During a severe winter in the US (1981-1982), a DC-8 airliner was delayed at Kennedy airport in New York (where the temperature was a freezing 2°F or minus 17°C) because mechanics needed to exchange a fuel pump (they received frost bite, which caused further delay). (Perrow, 1984, p. 135.)

After the aircraft finally got airborne after midnight, headed for San Francisco, passengers were told that there would be no coffee because the drinking water was frozen. Then the flight engineer discovered that he could not control the cabin pressure (which is held at a higher pressure than the thin air the aircraft is flying in so as to make the air breathable). Later investigation showed that the frozen drinking water had cracked the airplane's water tank. Heat from ducts to the tail section of the aircraft then melted the ice in the tank, and because of the crack in the tank, and the pressure in it, the newly melted water near the heat source sprayed

> out. It landed on the outflow valve that controls the cabin pressurization system (by allowing pressurized cabin air to vent outside). Once on the valve, the water turned to ice again because of the temperature of the outside air (minus 50°F or minus 45°C), which caused the valve to leak. The compressors for the cabin air could not keep up, leading to depressurization of the aircraft.
>
> The close proximity of parts that have no functional relationship, packed inside a compact airliner fuselage, can create the kind of interactive complexity and tight coupling that makes it hard to understand and control a propagating failure. Substituting broken parts was not possible (meaning tight coupling): the outflow valve is not reachable when airborne and a water tank cannot be easily replaced either (nor can a leak in it be easily fixed when airborne). The crew response to the pressurization problem, however, was rapid and effective – independent of their lack of understanding of the source of their pressurization problem. As trained, they got the airplane down to a breathable level in just three minutes and diverted to Denver for an uneventful landing there.

To Perrow, the two dimensions (interactive complexity and coupling) presented a serious dilemma. A system with high interactive complexity can only be effectively controlled by a decentralized organization. The reason is that highly interactive systems generate the sorts of non-routine situations that resist standardization (e.g., through procedures, which is a form of centralized control fed forward into the operation). Instead, the organization has to allow lower-level personnel considerable discretion and leeway to act as they see fit based on the situation, as well as encouraging direct interaction among lower-level personnel, so as to bring together the different kinds of expertise and perspective necessary to understand the problem.

A system with tight couplings, on the other hand, can in principle only be effectively controlled by a highly centralized organization, because tight coupling demands quick and coordinated responses. Disturbances that cascade through a system cannot be stopped quickly if a team with the right mix of expertise and backgrounds needs to be assembled first. Centralization, for example through procedures, emergency drills, or even automatic shut-downs or other machine interventions, is necessary to arrest such cascades quickly. Also, a conflict between different well-meaning interventions can make the situation worse, which means that activities oriented at arresting the failure propagation need to be extremely tightly coordinated.

To Perrow, an organization cannot be centralized and decentralized at the same time. So a dilemma arises if a system is both interactively complex and tightly coupled (e.g., nuclear power generation). A necessary conclusion for normal accidents theory is that systems that are both tightly coupled and interactively complex can therefore not be controlled effectively. This, however, is not the whole story. In the tightly coupled and interactively complex pressurization case above, the crew may not have been able to diagnose the source of the failure (which would indeed have involved decentralized multiple different perspectives, as well as access to various systems and components). Yet through centralization (procedures for dealing with pressurization problems are often trained, well-documented, brief and to the point) and extremely tight coordination (who does and says what in an emergency depressurization descent is very firmly controlled and goes unquestioned during execution of the task), the crew was able to stop the failure from propagating into a real disaster. Similarly, even if nuclear power plants are both interactively complex and tightly coupled, a mix of centralization and decentralization is applied so as to make propagating problems more manageable (e.g., thousands of pages of procedures and standard protocols exist, but so does the co-location of different kinds of expertise in one control room, to allow spontaneous interaction; and automatic shutdown sequences that get triggered in some situations can rule out the need for human intervention for up to 30 minutes).

NORMAL ACCIDENT THEORY AND "HUMAN ERROR"

At the sharp end of complex systems, normal accidents theory sees human error as a label for some of the effects of interactive complexity and tight coupling. Operators are the inheritors of a system that structurally conspires against their ability to make sense of what is going on and to recover from a developing failure. Investigations, infused with the wisdom of hindsight, says Perrow (1984) often turn up places where human operators should have zigged instead of zagged, as if that alone would have prevented the accident. Perrow invokes the idea of the fundamental surprise error when he comments on official inability to deal with the real structural nature of failure (e.g., through the investigations that are commissioned). The cause they find may sometimes be no more than the "cause" people are willing or able to afford. Indeed, to Perrow, the reliance on labels like "human error" has little to do with explanation and more with politics and power, something even formal or independent investigations are not always immune to:

> Formal accident investigations usually start with an assumption that the operator must have failed, and if this attribution can be made, that is the end of serious inquiry. Finding that faulty designs were responsible would entail enormous shutdown and retrofitting costs; finding that management was responsible would threaten those in charge, but finding that operators were responsible preserves the system, with some soporific injunctions about better training. (1984, p. 146)

Human error, in other words, can be a convenient and cheap label to use so as to control sunk costs and avoid having to upset elite interests. Behind the label, however, lie the real culprits: structural interactive complexity and tight coupling – features of risky technological systems such as nuclear power generation that society as a whole should be thinking critically about (Perrow, 1984).

That said, humans can hardly be the recipient victims of complexity and coupling alone. The very definition of Perrowian complexity actually involves both human and system, to the point where it becomes hard to see where one ends and the other begins. For example, interactions cannot be unfamiliar, unexpected, unplanned, or not immediately comprehensible in some system independent of the people who need to deal with them (and to whom they are either comprehensible or not). One hallmark of expertise, after all, is a reduction of the degrees of freedom that a decision presents to the problem-solver (Jagacinski and Flach, 2002), and an increasingly refined ability to recognize patterns of interactions and knowing what to do primed by such situational appreciation (Klein, Orasanu, and Calderwood, 1993). Perrowian complexity can thus not be a feature of a system by itself, but always has to be understood in relation to the people (and their expertise) who have to manage that system (e.g., Pew et al., 1981; Wagenaar and Groeneweg, 1987). This also means that the categories of complexity and coupling are not as independent as normal accident theory suggests.

Another problem arises when complexity and coupling are treated as stable properties of a system, because it misses the dynamic nature of much safety-critical work and the ebb and flow of cognitive and coordinative activity to manage it. During periods of crisis, or high demand, a system can become more difficult to control as couplings tighten and interactive complexity momentarily deepens. It renders otherwise visible interactions less transparent, less linear, creating interdependencies that are harder to understand and more difficult to correct. This can become especially problematic when important routines get interrupted, coordinated action breaks down and misunderstandings occur (Weick, 1990). The opposite goes too. Contractions in complexity and coupling can be met in centralized and de-centralized ways by people responsible for the safe operation of the system, creating new kinds of coordinated action and newly invented routines.

CASE 4.2 THE MAR KNOCKOUT CASE (COOK AND CONNOR, 2004)

During the Friday night shift in a large, tertiary care hospital, a nurse called the pharmacy technician on duty to report a problem with the medications just delivered for a ward patient in the unit dose cart. The call itself was not usual; occasionally there would be a problem with the medication delivered to the floor, especially if a new order was made after the unit dose fill list had been printed. In this case, however, the pharmacy had delivered medicines to the floor that had never been ordered for that patient. More importantly, the medicines that were delivered to the floor matched with the newly printed medication administration record (MAR). This was discovered during routine reconciliation of the previous day's MAR with the new one. The MAR that had just been delivered was substantially different from the one from the previous day but there was no indication in the patient's chart that these changes had been ordered. The pharmacy technician called up a computer screen that showed the patient's medication list. This list corresponded precisely to the new MAR and the medications that had been delivered to the ward.

While trying to understand what had happened to this patient's medication, the telephone rang again. It was a call from another ward where the nurses had discovered something wrong. For some patients, the unit dose cart contained drugs their patients were not taking, in others the cart did not contain drugs the patients were supposed to get. Other calls came in from other areas in the hospital, all describing the same situation. The problem seemed to be limited to the unit dose cart system; the intravenous medications were correct. In each case, the drugs that were delivered matched the newly printed MAR, but the MAR itself was wrong. The pharmacy technician notified the on-call pharmacist who realized that, whatever its source, the problem was hospital-wide. The MAR as a common mode created the kind of Perrowian complexity that made management of the problem extremely difficult: its consequences were showing up throughout the entire hospital, often in different guises and with different implications.

Consistent with normal accident theory, a technology that was introduced to improve safety, such as the dose checking software in this case, actually made it harder to achieve safety, for example, by making it difficult to upgrade to new software. Information technology makes it possible to perform work efficiently by speeding up much of the process. But the technology also makes it difficult to detect failures and recover from them. It introduces new forms of failure that are hard to appreciate before they occur. These failures are foreseeable but not foreseen. This was an event with system-wide consequences required decisive and immediate action to limit damage and potential damage. This action was expensive and potentially damaging to the prestige and authority of those who were in charge. The effective response required simultaneous, coordinated activity by experienced, skilled people.

Like many accidents, it was not immediately clear what had happened, only that something was wrong. It was now early Saturday morning and the pharmacy was confronting a crisis. First, the pharmacy computer system was somehow generating an inaccurate fill list. Neither MARs nor the unit dose carts already delivered to the wards could be trusted. There was no pharmacy computer-generated fill list that could be relied upon. Second, the wards were now without the right medications for the hospitalized patients and the morning medication administration process was about to begin. No one yet knew what was wrong with the pharmacy computer. Until it could be fixed, some sort of manual system was needed to provide the correct medications to the wards. Across the hospital, the unit dose carts were sent back to the pharmacy.

A senior pharmacist realized that the previous day's hard copy MARs as they were maintained on the wards were the most reliable available information about what medicines patients were supposed to receive. By copying the most recent MARs, the pharmacy could produce a manual fill list for each patient. For security reasons, there were no copying machines near the wards. There was a fax machine for each ward, however, and the pharmacy staff organized a ward-by-ward fax process to get hand-updated copies of each patient's MAR. Technicians used these faxes as manual fill lists to stock unit dose carts with correct medications. A decentralized response, in other words, that coordinated different kinds of expertise and background, making fortuitous use of substitutions (fax machines instead of copiers) helped people in the hospital manage the problem. A sudden contraction in interactive complexity through a common mode failure (MAR in this case) with a lack of centralized response capabilities (no central back-up) did not lead to total system breakdown because of the spontaneously organized response of practitioners throughout the system.

Ordinarily, MARs provided a way to track and reconcile the physician orders and medication administration process on the wards. In this instance they became the source of information about what medications were needed. Because the hospital did not yet have computer-based physician-order entry, copies of handwritten physician-orders were available. These allowed the satellite pharmacies to interact directly with the ward nurses to fill the gaps. Among the interesting features of the event was the absence of typewriters in the pharmacy. Typewriters, discarded years before in favor of computer-label printers, would have been useful for labeling medications. New technology displaces old technology, making it harder to recover from computer failures by reverting to manual operations.

The source of the failure remained unclear, as it often does, but that does not need to hamper the effectiveness of the coordinated response to it. There had been some problem with the pharmacy computer system during the previous evening. The pharmacy software detected a fault in the database integrity. The computer specialist had contacted the pharmacy software vendor and they had worked together through a fix to the problem. This fix proved unsuccessful so they reloaded a portion of the database from the most recent backup tape. After this reload, the system had appeared to work perfectly. The computer software had been purchased from a major vendor. After a devastating cancer chemotherapy accident in the institution, the software had been modified to include special dose-checking programs for chemotherapy. These modifications worked well but the pharmacy management had been slow to upgrade the main software package because it would require rewriting the dose-checking add-ons. Elaborate backup procedures were in place, including both frequent "change" backups and daily "full" backups onto magnetic tapes.

Working with the software company throughout the morning, the computer technicians were able to discover the reason that the computer system had failed. The backup tape was incomplete. Reloading had internally corrupted the database, and so the backup was corrupted because of a complex interlocking process related to the database management software that was used by the pharmacy application.

Under particular circumstances, tape backups could be incomplete in ways that remained hidden from the operator. The problem was not related to the fault for which the backup reloading was necessary. The immediate solution to the problem facing the pharmacy was to reload the last "full" backup (now over a day and a half old) and to re-enter all the orders made since that time. The many pharmacy technicians now collected all the handwritten order slips from the past 48 hours and began to enter these (the process was actually considerably more complex. For example, to bring the computer's view of the world up to date, its internal clock had to be set back, the prior day's fill list regenerated, the day's orders entered, the clock time set forward and the current day's morning fill list re-run). The manual system was used all Saturday. The computer system was restored by the end of the day. The managers and technicians examined the fill lists produced for the nightly fill closely and found no errors. The system was back "on-line".

As far as pharmacy and nursing management could determine, no medication misadministration occurred during this event. Some doses were delayed, although no serious consequences were identified. Several factors contributed to the hospital's ability to recover from the event. First, the accident occurred on a Friday night so that the staff had all day Saturday to recover and all day Sunday to observe the restored system for new failures. Few new patients are admitted on Saturday and the relatively slow tempo of operations allowed the staff to concentrate on recovering the system. Tight coupling, in other words, was averted fortuitously by the time of the week of the incident. Second, the hospital had a large staff of technicians and pharmacists who came in to restore operations. In addition, the close relationship between the software vendor and hospital information technical staff made it possible for the staff to diagnose the problem and devise a fix with little delay. The ability to quickly bring a large number of experts with operational experience together was critical to success, as normal accidents theory predicts is necessary in highly interactively complex situations. Third, the availability of the manual, paper records allowed these experts to "patch-up" the system and make it work in an unconventional but effective way. The paper MARs served as the basis for new fill lists and the paper copies of physician orders provided a "paper trail" that made it possible to replay the previous day's data entry, essentially fast forwarding the computer until it's "view" of the world was correct. Substitution of parts, in other words, was possible, thereby reducing coupling and arresting a cascade of failures. Fourth, the computer system and technical processes contributed. The backup process, while flawed in some ways, was essential to recovery: it provided the "full" backup needed. In other words, a redundancy existed that had not been deliberately designed-in (as is normally the case in tightly coupled systems according to normal accident theory).

The ability of organizations to protect themselves against system accidents (such as the MAR knockout close call) can, in worse cases than the one described above, fall victim to the very interactive complexity and tight coupling it must contain. Plans for emergencies, for example, are intended to help the organization deal with unexpected problems and developments for which are designed to be maximally persuasive to regulators, board members, surrounding communities, lawmakers and opponents of the technology, and as a result can become wildly unrealistic. Clarke and Perrow (1996) call them "fantasy documents," that fail to cover most possible accidents, lack any historical record that may function as a reality check, and are quickly based on obsolete contact details, organizational designs, function descriptions and divisions of responsibility. The problem with such fantasy documents is that they can function as an apparently legitimate placeholder that suggests that everything is under control. It inhibits the organization's commitment to continually reviewing and re-assessing its ability to deal with hazard. In other words, fantasy documents can impede organizational learning as well as organizational preparedness.

CONTROL THEORY

In response to the limitations of event chain models and their derivatives, such as the latent failure model, models based on control theory have been proposed for accident analysis instead. Accident models based on control theory explicitly look at accidents as emerging from interactions among system components. They usually do not identify single causal factors, but rather look at what may have gone wrong with the system's operation or organization of the hazardous technology that allowed an accident to take place. Safety, or risk management, is viewed as a control problem (Rasmussen, 1997), and accidents happen when component failures, external disruptions or interactions between layers and components are not adequately handled; when safety constraints that should have applied to the design and operation of the technology have loosened, or become badly monitored, managed, controlled. Control theory tries to capture these imperfect processes, which involve people, societal and organizational structures, engineering activities, and physical parts. It sees the complex interactions between those – as did man-made disaster theory – as eventually resulting in an accident (Leveson, 2002).

Control theory sees the operation of hazardous technology as a matter of keeping many interrelated components in a state of dynamic equilibrium (which means that control inputs, even if small, are continually necessary for the system to stay safe: it cannot be left on its own as could a statically stable system). Keeping a dynamically stable system in equilibrium happens through the use of feedback loops of information and control. Accidents are not the result of an initiating (root cause) event that triggers a series of events, which eventually leads to a loss. Instead, accidents result from interactions among components that violate the safety constraints on system design and operation, by which feedback and control inputs can grow increasingly at odds with the real problem or processes to be controlled. Unsurprisingly, concern with those control processes (how they evolve, adapt and erode) forms the heart of control theory as applied to organizational safety (Rasmussen, 1997; Leveson, 2002).

Degradation of the safety-control structure over time can be due to asynchronous evolution, where one part of a system changes without the related necessary changes in other parts. Changes to subsystems may have been carefully planned and executed in isolation, but consideration of their effects on other parts of the system, including the role they play in overall safety control, may remain neglected or inadequate. Asynchronous evolution can occur too when one part of a properly designed system deteriorates independent of other parts. In both cases, erroneous expectations of users or system components about the behavior of the changed or degraded subsystem may lead to accidents (Leveson, 2002). The more complex a system (and, by extension, the more complex its control structure), the more difficult it can become to map out the reverberations of changes (even carefully considered ones) throughout the rest of the system. Control theory embraces a much more complex idea of causation, taken from complexity theory. Small changes somewhere in the system, or small variations in the initial state of a process, can lead to huge consequences elsewhere. The Newtonian symmetry between cause and effect (still assumed in other models discussed in this chapter) no longer applies.

CASE 4.3 THE LEXINGTON COMAIR 5191 ACCIDENT (SEE NELSON, 2008)

Flight 5191 was a scheduled passenger flight from Lexington, Kentucky to Atlanta, Georgia, operated by Comair. On the morning of August 27, 2006, the Regional Jet that was being used for the flight crashed while attempting to take off. The aircraft was assigned runway 22 for the takeoff, but used runway 26 instead. Runway 26 was too short for a safe takeoff. The aircraft crashed just past the end of the runway, killing all 47 passengers and two of the three crew. The flight's first officer was the only survivor. At the time of the 5191 accident the LEX airport was in the final construction phases of a five year project. The First Officer had given the takeoff briefing and mentioned that "lights were out all over the place" (NTSB, 2007, p. 140) when he had flown in two nights before. He also gave the taxi briefing, indicating they would take taxiway Alpha to runway 22 and that it would be a short taxi. Unbeknownst to the crew, the airport signage was inconsistent with their airport diagram charts as a result of the construction. Various taxiway and runway lighting systems were out of operation at the time.

After a short taxi from the gate, the captain brought the aircraft to a stop short of runway 22, except, unbeknownst to him, they were actually short of runway 26. The control tower controller scanned runway 22 to assure there was no conflicting traffic, then cleared Comair 191 to take off. The view down runway 26 provided the illusion of some runway lights. By the time they approached the intersection of the two runways, the illusion was gone and the only light illuminating the runway was from the aircraft lights. This prompted the First Officer to comment "weird with no lights" and the captain responded "yeah" (NTSB, 2007, p. 157). During the next 14 seconds, they traveled the last 2,500 ft of remaining runway. In the last 100 feet of runway, the captain called "V1, Rotate, Whoa." The jet became momentarily airborne but then impacted a line of oak trees approximately 900 feet beyond the end of runway 26. From there, the aircraft erupted into flames and came to rest approximately 1,900 feet off the west end of runway 26.

Runway 26 was only 3,500 feet long and not intended for aircraft heavier than 12,000 pounds. Yet each runway had a crossing runway located approximately 1,500 feet from threshold. They both had an increase in elevation at the crossing runway. The opposite end of neither runway was visible during the commencement of the takeoff roll. Each runway had a dark-hole appearance at the end, and both had 150 foot wide pavement (runway 26 was edge striped to 75 feet). Neither runway had lighting down the center line, as that of runway 22 had been switched off as part of the construction (which the crew knew). Comair had no specified procedures to confirm compass heading with the runway. Modern Directional Gyros (DG) automatically compensate for precession, so it is no longer necessary to cross-check the DG with runway heading and compass indication. Many crews have abandoned the habit of checking this, as airlines have abandoned procedures for it. The 5191 crew was also fatigued, having accumulated sleep loss over the preceding duty period.

Comair had operated accident-free for almost 10 years when the 5191 accident occurred. During those 10 years, Comair approximately doubled its size, was purchased by Delta Air Lines Inc., became an all jet operator and, at the time of the 5191 accident, was in the midst of its first bankruptcy reorganization. As is typical with all bankruptcies, anything management believed was unnecessary was eliminated, and everything else was pushed to maximum utilization. In the weeks immediately preceding the 5191 accident, Comair had demanded large wage concessions from the pilots. Management had also indicated the possibility of furloughs and threatened to reduce the number of aircraft, thereby reducing the available flight hours and implying reduction of work force.

Data provided by Jeppesen, a major flight navigation and chart company, for NOTAM's (Notices to Airmen), did not contain accurate local information about the closure of taxiway Alpha North of runway 26. Comair, nor the crew, had any other way to get this information other than a radio broadcast at the airport itself, but there was no system in place for checking the completeness and accuracy of these either. According to the airport, the last phase of construction did not require a change in the route used to access runway 22; Taxiway A5 was simply renamed Taxiway A, but this change was not reflected on the crew's chart (indeed, asynchronous evolution). It would eventually become Taxiway A7.

> Several crews had acknowledged difficulty dealing with the confusing aspects of the north end taxi operations to runway 22, following the changes which affected a seven-day period prior to the 5191 accident. One captain, who flew in and out of LEX numerous times a month, stated that after the changes "there was not any clarification about the split between old alpha taxiway and the new alpha taxiway and it was confusing." A First Officer, who also regularly flew in and out of LEX, expressed that on their first taxi after the above changes, he and his captain "were totally surprised that taxiway Alpha was closed between runway 26 and runway 22." The week before, he used taxiway Alpha (old Alpha) to taxi all the way to runway 22. It "was an extremely tight area around runway 26 and runway 22 and the chart did not do it justice." Even though these and, undoubtedly, other instances of crew confusion occurred during the seven-day period of August 20–27, 2006, there were no effective communication channels to provide this information to LEX, or anyone else in the system. After the 5191 accident, a small group of aircraft maintenance workers expressed concern that they, too, had experienced confusion when taxiing to conduct engine run-up's. They were worried that an accident could happen, but did not know how to effectively notify people who could make a difference.
>
> The regulator had not approved the publishing of interim airport charts that would have revealed the true nature of the situation. It had concluded that changing the chart over multiple revision cycles would create a high propensity for inaccuracies to occur, and that, because of the multiple chart changes, the possibilities for pilot confusion would be magnified.

Control theory has part of its background in control engineering, which helps the design of control and safety systems in hazardous industrial or other processes, particularly with software applications (e.g., Leveson and Turner, 1993). The models, as applied to organizational safety, are concerned with how a lack of control allows a migration of organizational activities towards the boundary of acceptable performance, and there are several ways to represent the mechanisms by which this occurs. Systems dynamics modeling does not see an organization as a static design of components or layers. It readily accepts that a system is more than the sum of its constituent elements. Instead, they see an organization as a set of constantly changing and adaptive processes focused on achieving the organization's multiple goals and adapting around its multiple constraints. The relevant units of analysis in control theory are therefore not components or their breakage (e.g., holes in layers of defense), but system constraints and objectives (Rasmussen, 1997; Leveson, 2002):

> Human behavior in any work system is shaped by objectives and constraints which must be respected by the actors for work performance to be successful. Aiming at such productive targets, however, many degrees of freedom are left open which will have to

COMPLEXITY, CONTROL AND SOCIOLOGICAL MODELS 73

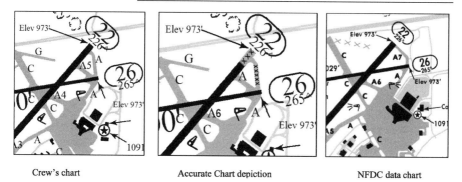

Figure 4.1   The difference between the crew's chart on the morning of the accident, the actual situation (center) and the eventual result of the reconstruction (NFDC or National Flight Data Center chart to the right). From Nelson, 2008

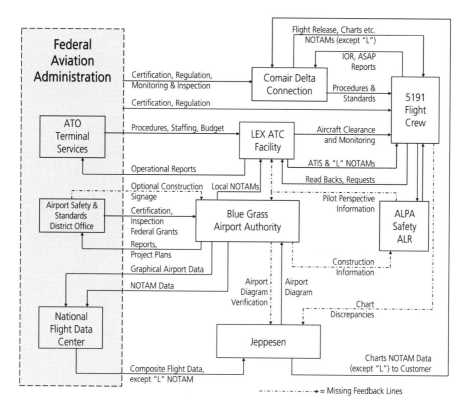

Figure 4.2   The structure responsible for safety-control during airport construction at Lexington, and how control deteriorated. Lines going into the left of a box represent control actions, lines from the top or bottom represent feedback

be closed by the individual actor by an adaptive search guided by process criteria such as workload, cost effectiveness, risk of failure, joy of exploration, and so on. The work space within which the human actors can navigate freely during this search is bounded by administrative, functional and safety-related constraints. The normal changes found in local work conditions lead to frequent modifications of strategies and activity will show great variability … During the adaptive search the actors have ample opportunity to identify 'an effort gradient' and management will normally supply an effective 'cost gradient'. The result will very likely be a systematic migration toward the boundary of functionally acceptable performance and, if crossing the boundary is irreversible, an error or an accident may occur. (Rasmussen, 1997, p. 189)

The dynamic interplay between these different constraints and objectives is illustrated in Figure 4.3.

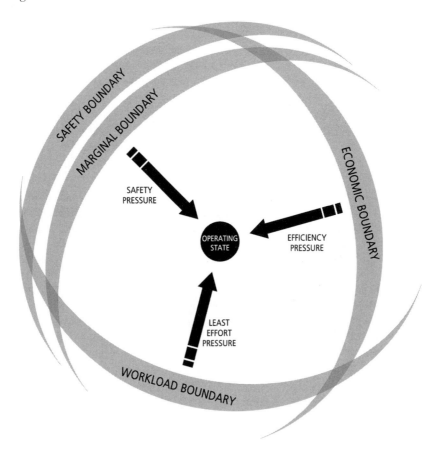

Figure 4.3 A space of possible organizational action is bounded by three constraints: safety, workload and economics. Multiple pressures act to move the operating point of the organization in different directions. (Modified from Cook and Rasmussen, 2005)

CONTROL THEORY AND "HUMAN ERROR"

Control theory sees accidents as the result of normal system behavior, as organizations try to adapt to the multiple, normal pressures that operate on it every day. Reserving a place for "inadequate" control actions, as some models do, of course does re-introduce human error under a new label (accidents are not the result of human error, but the result of inadequate control – what exactly is the difference then?). Systems dynamics modeling must deal with that problem by recursively modeling the constraints and objectives that govern the control actions at various hierarchical levels, thereby explaining the "inadequacy" as a normal result of normal pressures and constraints operating on that level from above and below, and in turn influencing the objectives and constraints for other levels. Rasmussen (1997) does this by depicting control of a hazardous technology as a nested series of reciprocally constraining hierarchical levels, down from the political and governmental level, through regulators, companies, management, staff, all the way to sharp-end workers. This nested control structure is also acknowledged by Leveson (2002).

In general, systems dynamics modeling is not concerned with individual unsafe acts or errors, or even individual events that may have helped trigger an accident sequence. Such a focus does not help, after all, in identifying broader ways to protect the system against similar migrations towards risk in the future. Systems dynamics modeling also rejects the depiction of accidents in the traditionally physical way as the latent failure model does, for example. Accidents are not about particles, paths of traveling or events of collision between hazard and process-to-be-protected (Rasmussen, 1997). The reason for rejecting such language (even visually) is that removing individual unsafe acts, errors or singular events from a presumed or actual accident sequence only creates more space for new ones to appear if the same kinds of systemic constraints and objectives are left similarly ill-controlled in the future. The focus of control theory is therefore not on erroneous actions or violations, but on the mechanisms that help generate such behaviors at a higher level of functional abstraction – mechanisms that turn these behaviors into normal, acceptable and even indispensable aspects of an actual, dynamic, daily work context.

Fighting violations or other deviations from presumed ways of operating safely – as implicitly encouraged by other models discussed above – is not very useful according to control theory. A much more effective strategy for controlling behavior is by making the boundaries of system performance explicit and known, and to help people develop skills at coping with the edges of those boundaries. Ways proposed by Rasmussen (1997) include increasing the margin from normal operation to the loss-of-control boundary. This, however, is only partially effective because of risk homeostasis and the law of stretched systems – the tendency for a system under goal pressures to gravitate back to a certain level of risk acceptance, even after interventions to make it safer. In other words, if the boundary of safe operations is moved further away, then normal operations will likely follow not long after – under pressure, as they always are, from the objectives of efficiency and less effort.

CASE 4.4   RISK HOMEOSTASIS

One example of risk homeostasis is the introduction of anti-lock brakes and center-mounted brake lights on cars. Both these interventions serve to push the boundary of safe operations further out, enlarging the space in which driving can be done safely (by notifying drivers better when a preceding vehicle brakes, and by improving the vehicle's own braking performance independent of road conditions). However, this gain is eaten up by the other pressures that push on the operating point: drivers will compensate by closing the distance between them and the car in front (after all, they can see better when it brakes now, and they may feel their own braking performance has improved). The distance between the operating point and the boundary of safe operations closes up`.

Another way is to increase people's awareness that the system may be drifting towards the boundary, and then launching safety campaign to push back in the opposite direction (Rasmussen, 1997).

CASE 4.5   TAKE-OFF CHECKLISTS AND THE PRESSURE TO DEPART ON-TIME

Airlines frequently struggle with on-time performance, particularly in heavily congested parts of the world, where so-called slot times govern when aircraft may become airborne. Making a slot time is critical, as it can be hours for a new slot to open up if the first one is missed. This push for speed can lead to problems with for example pre-take off checklists, and airlines regularly have problems with attempted take-offs in airplanes that are not correctly configured (particularly the wing flaps which help the aircraft fly at slower speeds such as in take-off and landing).

One airline published a flight safety news letter that was distributed to all its pilots. The letter counted seven such configuration events in half a year, where aircraft did not have wing flaps selected before taking off, even when the item "flaps" on the before take-off checklist was read and responded to by the pilots. Citing no change in procedures (so that could not be the explanation), the safety letter went on to speculate whether stress or complacency could be a factor, particularly as it related to the on-time performance goals (which are explicitly stated by the airline elsewhere). Slot times played a role in almost half the events. While acknowledging that slot times and on-time performance were indeed important goals for the airline, the letter went on to say that flight safety should not be sacrificed for those goals. In an attempt to help crews

> develop their skills at coping with the boundaries, the letter also suggested that crew members should act on 'gut' feelings and speak out loudly as soon as something was detected that was amiss, particularly in high workload situations.

Leaving both pressures in place (a push for greater efficiency and a safety campaign pressing in the opposite direction) does little to help operational people (pilots in the case above) cope with the actual dilemma at the boundary. Also, a reminder to try harder and watch out better, particularly during times of high workload, is a poor substitute for actually developing skills to cope at the boundary. Raising awareness, however, can be meaningful in the absence of other possibilities for safety intervention, even if the effects of such campaigns tend to wear off quickly. Greater safety returns can be expected only if something more fundamental changes in the behavior-shaping conditions or the particular process environment (e.g., less traffic due to industry slow-down, leading to less congestion and fewer slot times). In this sense, it is important to raise awareness about the migration toward boundaries throughout the organization, at various managerial levels, so that a fuller range of countermeasures is available beyond telling front-line operators to be more careful. Organizations that are able to do this effectively have sometimes been dubbed high-reliability organizations.

HIGH-RELIABILITY THEORY

High reliability theory describes the extent and nature of the effort that people, at all levels in an organization, have to engage in to ensure consistently safe operations despite its inherent complexity and risks. Through a series of empirical studies, high-reliability organizational (HRO) researchers found that through leadership safety objectives, the maintenance of relatively closed systems, functional decentralization, the creation of a safety culture, redundancy of equipment and personnel, and systematic learning, organizations could achieve the consistency and stability required to effect failure-free operations (LaPorte and Consolini, 1991). Some of these categories were very much inspired by the worlds studied – naval aircraft carriers, for example (Rochlin, LaPorte and Roberts, 1987). There, in a relatively self-contained and disconnected closed system, systematic learning was an automatic by-product of the swift rotations of naval personnel, turning everybody into instructor and trainee, often at the same time. Functional decentralization meant that complex activities (like landing an aircraft and arresting it with the wire at the correct tension) were decomposed into simpler and relatively homogenous tasks, delegated down into small workgroups with substantial autonomy to intervene and stop the entire process independent of rank. HRO researchers found many forms of redundancy – in technical systems, supplies, even decision-making and management hierarchies, the latter through shadow units and multi-skilling.

When HRO researchers first set out to examine how safety is created and maintained in such complex systems, they focused on errors and other negative indicators, such as incidents, assuming that these were the basic units that people in these organizations used to map the physical and dynamic safety properties of their production technologies, ultimately to control risk (Rochlin, 1999). The assumption was wrong: they were not. Operational people, those who work at the sharp end of an organization, hardly defined safety in terms of risk management or error avoidance. Ensuing empirical work by HRO, stretching across decades and a multitude of high-hazard, complex domains (aviation, nuclear power, utility grid management, navy) would paint a more complex picture. Operational safety – how it is created, maintained, discussed, mythologized – is much more than the control of negatives. As Rochlin (1999, p. 1549) put it:

> The culture of safety that was observed is a dynamic, intersubjectively constructed belief in the possibility of continued operational safety, instantiated by experience with anticipation of events that could have led to serious errors, and complemented by the continuing expectation of future surprise.

The creation of safety, in other words, involves a belief about the possibility to continue operating safely. This belief is built up and shared among those who do the work every day. It is moderated or even held up in part by the constant preparation for future surprise – preparation for situations that may challenge people's current assumptions about what makes their operation risky or safe. It is a belief punctuated by encounters with risk, but it can become sluggish by overconfidence in past results, blunted by organizational smothering of minority viewpoints, and squelched by acute performance demands or production concerns. But that also makes it a belief that is, in principle, open to organizational or even regulatory intervention so as to keep it curious, open-minded, complexly sensitized, inviting of doubt, and ambivalent toward the past (e.g., Weick, 1993).

HIGH RELIABILITY AND "HUMAN ERROR"

An important point for the role of "human error" in high reliability theory is that safety is not the same as reliability. A part can be reliable, but in and of itself it can't be safe. It can perform its stated function to the expected level or amount, but it is context, the context of other parts, of the dynamics and the interactions and cross-adaptations between parts, that make things safe or unsafe. Reliability as an engineering property is expressed as a component's failure rate over a period of time. In other words, it addresses the question of whether a component lives up to its pre-specified performance criteria. Organizationally, reliability is often associated with a reduction in variability, and an increase in replicability: the same process, narrowly guarded, produces the same predictable outcomes. Becoming highly reliable may be a desirable goal for unsafe or moderately safe operations (Amalberti, 2001). The guaranteed production of standard outcomes through consistent component performance is a way to reduce failure probability in those operations, and it is often expressed as a drive to eliminate "human errors" and technical breakdowns.

In moderately safe systems, such as chemical industries or driving or chartered flights, approaches based on reliability can still generate significant safety returns (Amalberti,

2001). Regulations and safety procedures have a way of converging practice onto a common basis of proven performance. Collecting stories about negative near-miss events (errors, incidents) has the benefit in that the same encounters with risk show up in real accidents that happen to that system. There is, in other words, an overlap between the ingredients of incidents and the ingredients of accidents: recombining incident narratives has predictive (and potentially preventive) value. Finally, developing error-resistant and error-tolerant designs helps cut down on the number of errors and incidents.

The monitoring of performance through operational safety audits, error counting, process data collection, and incident tabulations has become institutionalized and in many cases required by legislation or regulation. As long as an industry can assure that components (parts, people, companies, countries) can comply with pre-specified and auditable criteria, it affords the belief that it has a safe system. Quality assurance and safety management within an industry are often mentioned in the same sentence or used under one department heading. The relationship is taken as non-problematic or even coincident. Quality assurance is seen as a fundamental activity in risk management. Good quality management will help ensure safety.

Such beliefs may well have been sustained by models such as the latent failure model discussed above, which posited that accidents are the result of a concatenation of factors, a combination of active failures at the sharp end with latent failures from the blunt end (the organizational, regulatory, societal part) of an organization. Accidents represent opportunistic trajectories through imperfectly sealed or guarded barriers that had been erected at various levels (procedural, managerial, regulatory) against them. This structuralist notion plays into the hand of reliability: the layers of defense (components) should be checked for their gaps and holes (failures) so as to guarantee reliable performance under a wide variety of conditions (the various line-ups of the layers with holes and gaps). People should not violate rules, process parameters should not exceed particular limits, acme nuts should not wear beyond this or that thread, a safety management system should be adequately documented, and so forth.

This model also sustains decomposition assumptions that are not really applicable to complex systems (see Leveson, 2002). For example, it suggests that each component or sub-system (layer of defense) operates reasonably independently, so that the results of a safety analysis (e.g., inspection or certification of people or components or sub-systems) are not distorted when we start putting the pieces back together again. It also assumes that the principles that govern the assembly of the entire system from its constituent sub-systems or components is straightforward. And that the interactions, if any, between the sub-systems will be linear: not subject to unanticipated feedback loops or non-linear interactions.

The assumptions baked into that reliability approach mean that aviation should continue to strive for systems with high theoretical performance and a high safety potential – that the systems it designs and certifies are essentially safe, but that they are undermined by technical breakdowns and human errors. The elimination of this residual reliability "noise" is still a widely-pursued goal, as if industries are the custodian of an already safe system that merely needs protection from unpredictable, erratic components that are the remaining sources of unreliability. This common sense approach, says Amalberti (2001), which indeed may have helped some systems progress to their safety levels of today, is beginning to lose its traction. This is echoed by Vaughan (1996, p. 416):

> We should be extremely sensitive to the limitations of known remedies. While good management and organizational design may reduce accidents in certain systems, they can never prevent them ... technical system failures may be more difficult to avoid than even the most pessimistic among us would have believed. The effect of unacknowledged and invisible social forces on information, interpretation, knowledge, and – ultimately – action, are very difficult to identify and to control.

Many systems, even after progressing beyond being moderately safe, are still embracing this notion of reliability with vigor – not just to maintain their current safety level (which would logically be non-problematic, in fact, it would even be necessary) but also as a basis for increasing safety even further. But as progress on safety in more mature systems (e.g., commercial aviation) has become asymptotic, further optimization of this approach is not likely to generate significant safety returns. In fact, there could be indications that continued linear extensions of a traditional-componential reliability approach could paradoxically help produce a new kind of system accident at the border of almost totally safe practice (Amalberti, 2001, p. 110):

> The safety of these systems becomes asymptotic around a mythical frontier, placed somewhere around 5×10^{-7} risks of disastrous accident per safety unit in the system. As of today, no man-machine system has ever crossed this frontier, in fact, solutions now designed tend to have devious effects when systems border total safety.

The accident described below illustrates how the reductionist reliability model applied to understanding safety and risk (taking systems apart and checking whether individual components meet prespecified criteria) may no longer work well, and may in fact have contributed to the accident. Through a concurrence of functions and events, of which a language barrier was a product as well as constitutive, the flight of a Boeing 737 out of Cyprus in 2005 may have been pushed past the edge of chaos, into that area in non-linear dynamics where new system behaviors emerge that cannot be anticipated using reductive logic, and negate the Newtonian assumption of symmetry between cause and consequence.

CASE 4.6   HELIOS AIRWAYS B737, AUGUST 2005

On 13 August 2005, on the flight prior to the accident, a Helios Airways Boeing 737-300 flew from London to Larnaca, Cyprus. The cabin crew noted a problem with one of the doors, and convinced the flight crew to write that the "Aft service door requires full inspection" in the aircraft logbook. Once in Larnaca, a ground engineer performed an inspection of the door and carried out a cabin pressurization-leak check during the night. He found no defects. The aircraft was released from maintenance at 03:15 and scheduled for flight 522 at 06:00 via Athens, Greece to Prague, Czech Republic (AAISASB, 2006).

A few minutes after taking off from Larnaca, the captain called the company in Cyprus on the radio to report a problem with his equipment cooling and the take-off configuration horn (which warns pilots that the aircraft is not configured properly for take-off, even though it evidently had taken off successfully already). A ground engineer was called to talk with the captain, the same ground engineer who had worked on the aircraft in the night hours before. The ground engineer may have suspected that the pressurization switches could be in play (given that he had just worked on the aircraft's pressurization system), but his suggestion to that effect to the captain was not acted on. Instead, the captain wanted to know where the circuit breakers for his equipment cooling were so that he could pull and reset them.

During this conversation, the oxygen masks deployed in the passenger cabin as they are designed to do when cabin altitude exceeds 14,000 feet. The conversation with the ground engineer ended, and would be the last that would have been heard from flight 522. Hours later, the aircraft finally ran out of fuel and crashed in hilly terrain north of Athens. Everybody on board had been dead for hours, except for one cabin attendant who held a commercial pilots license. Probably using medical oxygen bottles to survive, he finally had made it into the cockpit, but his efforts to save the aircraft were too late. The pressurization system had been set to manual so that the engineer could carry out the leak check. It had never been set back to automatic (which is done in the cockpit), which meant the aircraft did not pressurize during its ascent, unless a pilot had manually controlled the pressurization outflow valve during the entire climb. Passenger oxygen had been available for no more than 15 minutes, the captain had left his seat, and the co-pilot had not put on an oxygen mask.

Helios 522 is unsettling and illustrative, because nothing was "wrong" with the components. They all met their applicable criteria. "The captain and First Officer were licensed and qualified in accordance with applicable regulations and Operator requirements. Their duty time, flight time, rest time, and duty activity patterns were according to regulations. The cabin attendants were trained and qualified to perform their duties in accordance with existing requirements" (AAISASB, 2006, p. 112). Moreover, both pilots had been declared medically fit, even though postmortems revealed significant arterial clogging that may have accelerated the effects of hypoxia. And while there are variations in what JAR-compliant means as one travels across Europe, the Cypriot regulator (Cyprus DCA, or Department of Civil Aviation) complied with the standards in JAR OPS 1 and Part 145. This was seen to with help from the UK CAA, who provided inspectors for flight operations and airworthiness audits by means of contracts with the DCA. Helios and the maintenance organization were both certified by the DCA.

The German captain and the Cypriot co-pilot met the criteria set for their jobs. Even when it came to English, they passed. They were within the bandwidth of quality control within which we think system safety is guaranteed, or at least highly likely. That layer of defense – if you choose speak that language – had no holes as far as our system for checking and regulation could determine in advance. And we thought we could line these sub-systems up linearly, without complicated interactions. A German captain, backed up by a Cypriot co-pilot. In a long-since certified airframe, maintained by an approved organization. The assembly of the total system could not be simpler. And it must have, should have, been safe.

Yet the brittleness of having individual components meet prespecified criteria became apparent when compounding problems pushed demands for crew coordination beyond the routine. As the AAISASB observed, "Sufficient ease of use of English for the performance of duties in the course of a normal, routine flight does not necessarily imply that communication in the stress and time pressure of an abnormal situation is equally effective. The abnormal situation can potentially require words that are not part of the 'normal' vocabulary (words and technical terms one used in a foreign tongue under normal circumstances), thus potentially leaving two pilots unable to express themselves clearly. Also, human performance, and particularly memory, is known to suffer from the effects of stress, thus implying that in a stressful situation the search and choice of words to express one's concern in a non-native language can be severely compromise … In particular, there were difficulties due to the fact that the captain spoke with a German accent and could not be understood by the British engineer. The British engineer did not confirm this, but did claim that he was also unable to understand the nature of the problem that the captain was encountering." (pp. 122–123).

The irony is that the regulatory system designed to standardize aviation safety across Europe, has, through its harmonization of crew licensing, also legalized the blending of a large number of crew cultures and languages inside of a single airliner, from Greek to Norwegian, from Slovenian to Dutch. On the 14th of August 2005, this certified and certifiable system was not able to recognize, adapt to, and absorb a disruption that fell outside the set of disturbances it was designed to handle. The "stochastic fit" (see Snook, 2000) that put together this crew, this engineer, from this airline, in this airframe, with these system anomalies, on this day, outsmarted how we all have learned to create and maintain safety in an already very safe industry. Helios 522 testifies that the quality of individual components or sub-systems predicts little about how they can stochastically and non-linearly recombine to outwit our best efforts at anticipating pathways to failure.

5
RESILIENCE ENGINEERING

Resilience Engineering represents a way of thinking about safety that departs from conventional risk management approaches (e.g., error tabulation, violations, calculation of failure probabilities). Furthermore, it looks for ways to enhance the ability of organizations to monitor and revise risk models, to create processes that are robust yet flexible, and to use resources proactively in the face of disruptions or ongoing production and economic pressures. Accidents, according to Resilience Engineering, do not represent a breakdown or malfunctioning of normal system functions, but rather represent the breakdowns in the adaptations necessary to cope with the real world complexity. As control theory suggested with its emphasis on dynamic stability, individuals and organizations must always adjust their performance to current conditions; and because resources and time are finite it is inevitable that such adjustments are approximate. Success has been ascribed to the ability of groups, individuals, and organizations to anticipate the changing shape of risk before damage occurs; failure is the temporary or permanent absence of that ability.

CASE 5.1 NASA ORGANIZATIONAL DRIFT INTO THE *COLUMBIA* ACCIDENT

While the final breakup sequence of the space shuttle *Columbia* could be captured by a sequence-of-events model, the organizational background behind it takes a whole different form of analysis, and has formed a rich trove of inspiration for thinking about how to engineer resilience into organizations.

A critical precursor to the mission was the re-classification of foam events from in-flight anomalies to maintenance and turn-around issues, something that significantly degraded the safety status of foam strikes. Foam loss was increasingly seen as an accepted risk or even, as one pre-launch briefing put it, "not a safety of flight issue" (CAIB, 2003, p. 126).

This shift in the status of foam events is an important part of explaining the limited and fragmented evaluation of the *Columbia* foam strike and how analysis of that foam event never reached the problem-solving groups that were practiced at investigating anomalies, their significance and consequences, that is, Mission Control.

What was behind this reclassification, how could it make sense for the organization at the time? Pressure on schedule issues produced a mindset centered on production goals. There are several ways in which this could have played a role: schedule pressure magnifies the importance of activities that affect turnaround; when events are classified as in-flight anomalies a variety of formal work steps and checks are invoked; the work to assess anomalies diverts resources from the tasks to be accomplished to meet turnaround pressures. In fact the rationale for the reclassification was quite weak, and flawed. The CAIB's examination reveals that no cross-checks were in place to detect, question, or challenge the specific flaws in the rationale. Managers used what on the surface looked like technical analyses to justify previously reached conclusions, rather than the robust cognitive process of using technical analyses to test tentative hypotheses.

It would be very important to know more about the mindset and stance of different groups toward this shift in classification. For example, one would want to consider: Was the shift due to the salience of the need to improve maintenance and turnaround? Was this an organizational structure issue (which organization focuses on what aspects of problems)? What was Mission Control's reaction to the reclassification? Was it heard about by other groups? Did reactions to this shift remain underground relative to formal channels of communication?

Interestingly, the organization had three categories of risk: in-flight anomalies, accepted risks, and non-safety issues. As the organization began to view foam events as an accepted risk, there was no formal means for follow-up with a re-evaluation of an "accepted" risk to assess if it was in fact acceptable as new evidence built up or as situations changed. For all practical purposes, there was no difference between how the organization was handling non-safety issues and how it was handling accepted risks (i.e., accepted risks were being thought of and acted on no differently than non-safety issues). Yet the organization acted as if items placed in the accepted risk category were being evaluated and handled appropriately (i.e., as if the assessment of the hazard was accurate and up to date and as if the countermeasures deployed were still shown to be effective).

Foam events were only one source of debris strikes that threaten different aspects of the orbiter structure. Debris strikes carry very different risks depending on where and what they strike. The hinge in considering the response to the foam strike on STS-107 is that the debris struck the leading-edge structure (RCC panels and seals) and not the tiles. Did concern and progress on improving tiles block the ability to see risks to other structures? Did NASA regard the leading edge as much less vulnerable to damage than tiles? This is important because the damage in a previous mission (STS-45) provided an opportunity to focus on the leading-edge structure and reconsider the margins to failure of that structure given strikes by various kinds of debris. Did this mission create a sense that the leading-edge structure was less vulnerable than tiles? Did this mission fail to revise a widely held belief that the RCC leading-edge panels were more robust to debris strikes than they really were? Who followed up the damage to the RCC panel and what did they conclude? Who received the results? How were risks to non-tile structures evaluated and considered – including landing gear door structures? More information about the follow-up to leading-edge damage in STS-45 would shed light on how this opportunity was missed.

A management stance emerged early in the *Columbia* mission which downplayed significance of the strike. The initial and very preliminary assessments of the foam strike created a stance toward further analysis that this was not a critical or important issue for the mission. The stance developed and took hold before there were results from any technical analyses. This indicates that preliminary judgments were biasing data evaluation, instead of following a proper engineering evaluation process where data evaluation points teams and management to conclusions.

Indications that the event was outside of boundary conditions for NASA's understanding of the risks of debris strikes seemed to go unrecognized. When events fall outside of boundaries of past data and analysis tools and when the data available includes large uncertainties, the event is by definition anomalous and of high risk. While personnel noted the specific indications in themselves, no one was able to use these indicators to trigger any deeper or wider recognition of the nature of the anomaly in this situation. This pattern of seeing the details but being unable to recognize the big picture is commonplace in accidents.

As the Debris Assessment Team (DAT) was formed after the strike was detected and began to work, the question arose: "Is the size of the debris strike 'out-of-family' or 'in-family' given past experience?" While the team looked at past experience, it was unable to get a consistent or informative read on how past events indicated risk for this event. It appears no other groups or representatives of other technical areas were brought into the picture. This absence of any cross-checks is quite notable and inconsistent with how Mission Control groups evaluate in-flight anomalies. Past studies indicate that a review or interaction with another group would have provided broadening checks which help uncover inconsistencies and gaps as people need to focus their analysis, conclusions, and justifications for consideration and discussion with others.

Evidence that the strike posed a risk of serious damage kept being encountered – RCC panel impacts at angles greater than 15 degrees predicted coating penetration (CAIB, 2003, p. 145), foam piece 600 times larger than ice debris previously analyzed (CAIB, 2003, p. 143), models predicting tile damage deeper than tile thickness (CAIB, 2003, p. 143). Yet a process of discounting evidence discrepant with the current assessment went on several times (though eventually the DAT concerns seem to focus on the landing gear doors rather than the leading-edge structure).

Given the concerns about potential damage that arose in the DAT and given its desire to determine the location more definitively, the question arises: did the team conduct contingency analyses of damage and consequences across the different candidates sites – leading edge, landing gear door seals, tiles? Based on the evidence compiled in the CAIB report, there was no contingency analysis or follow through on the consequences if the leading-edge structure (RCC) was the site damaged. This is quite puzzling as this was the team's first assessment of location and in hindsight their initial estimate proved to be reasonably accurate.

This lack of follow-through, coupled with the DAT's growing concerns about the landing gear door seals, seems to indicate that the team may have viewed the leading-edge structures as more robust to strikes than other orbiter structures. The CAIB report fails to provide critical information about how different groups viewed the robustness or vulnerability of the leading-edge structure to damage from debris strikes (of course, post-accident these beliefs can be quite hard to determine, but various memos/analyses may indicate more about the perception risks to this part of the orbiter). Insufficient data is available to understand why RCC damage was under-pursued by the Debris Assessment Team.

There was a fragmented view of what was known about the strike and its potential implications over time, people, and groups. There was no place, artifact, or person who had a complete and coherent view of the analysis of the foam strike event (note a coherent view includes understanding the gaps and uncertainties in the data or analysis to that point). This contrasts dramatically with how Mission Control works to investigate and handle anomalies where there are clear lines of responsibility to have a complete, coherent view of the evolving analysis vested in the relevant flight controllers and in the flight director. Mission Control has mechanisms to keep different people in the loop (via monitoring voice loops, for example) so that all are up to date on the current picture of situation. Mission Control also has mechanisms for correcting assessments as analysis proceeds, whereas in this case the fragmentation and partial views seemed to block reassessment and freeze the organization in an erroneous assessment. As the DAT worked at the margins of knowledge and data, its partial assessments did not benefit from cross-checks through interactions with other technical groups with different backgrounds and assumptions. There is no report of a technical review process that accompanied its work. Interactions with people or groups with different knowledge and assumptions is one of the best ways to improve assessments and to aid revision of assessments. Mission Control anomaly-response includes many opportunities for cross-checks to occur. In general, it is quite remarkable that the groups practiced at anomaly response – Mission Control – never became involved in the process.

The process of analyzing the foam strike by the DAT broke down in many ways. The fact that this group also advocated steps that we now know would have been valuable (the request for imagery to locate the site of the foam strike) leads us to miss the generally fragmented distributed problem-solving process. The fragmentation also occurred across organizational levels (DAT to Mission Management Team (MMT)). Effective collaborative problem-solving requires more direct participation by members of the analysis team in the overall decision-making process. This is not sufficient of course; for example, the MMT's stance already defined the situation as, "Show me that the foam strike is an issue" rather than "Convince me the anomaly requires no response or contingencies." Overall, the evidence points to a broken distributed problem-solving process – playing out in between organizational boundaries. The fragmentation in this case indicates the need for a senior technical focal point to integrate and guide the anomaly analysis process (e.g., the flight director role). And this role requires real authority. The MMT and the MMT chair were in principle in a position to supply this role, but: Was the MMT practiced at providing the integrative problem-solving role? Were there other cases where significant analysis for in flight anomalies was guided by the MMT or were they all handled by the Mission Control team? The problem-solving process in this case has the odd quality of being stuck in limbo: not dismissed or discounted completely, yet unable to get traction as an in-flight anomaly to be thoroughly investigated with contingency analyses and re-planning activities. The dynamic appears to be a management stance that puts the event outside of safety of flight (e.g., conclusions drove, or eliminated, the need for analysis and investigation, rather than investigations building the evidence from which one would draw conclusions). Plus, the DAT exhibited a fragmented problem-solving process that failed to integrate partial and uncertain data to generate a big picture – that is, the situation was outside the understood risk boundaries and carried significant uncertainties.

The *Columbia* case reveals a number of classic patterns that have helped shape the ideas behind resilience engineering – some of these patterns have part of their basis in the earlier models described in this chapter:

- Drift toward failure as defenses erode in the face of production pressure.
- An organization that takes past success as a reason for confidence instead of investing in anticipating the changing potential for failure.
- Fragmented distributed problem-solving process that clouds the big picture.
- Failure to revise assessments as new evidence accumulates.
- Breakdowns at the boundaries of organizational units that impede communication and coordination.

The *Columbia* case provides an example of a tight squeeze on production goals, which created strong incentives to downplay schedule disruptions. With shrinking time/resources available, safety margins were likewise shrinking in ways which the organization couldn't see. Goal tradeoffs often proceed gradually as pressure leads to a narrowing of focus on some goals while obscuring the tradeoff with other goals. This process usually happens when acute goals like production/efficiency take precedence over chronic goals like safety. The dilemma of production/safety conflicts is this: if organizations never sacrifice production pressure to follow up warning signs, they are acting much too riskily. On the other hand, if uncertain "warning" signs always lead to sacrifices on acute goals, can the organization operate within reasonable parameters or stakeholder demands? It is precisely at points of intensifying production pressure that extra safety investments need to be made in the form of proactive searching for side-effects of the production pressure and in the form or reassessing the risk space – safety investments are most important when least affordable. This raises the following questions:

- How does a safety organization monitor for drift and its associated signs, in particular, a means to recognize when the side-effects of production pressure may be increasing safety risks?
- What indicators should be used to monitor the organization's model of itself, how it is vulnerable to failure, and the potential effectiveness of the countermeasures it has adopted?
- How does production pressure create or exacerbate tradeoffs between some goals and chronic concerns like safety?
- How can an organization add investment in safety issues at the very time when the organization is most squeezed? For example, how does an organization note a reduction in margins and follow through by rebuilding margin to boundary conditions in new ways?

Another general pattern identified in *Columbia* is that an organization takes past success as a reason for confidence instead of digging deeper to see underlying risks. During the drift toward failure leading to the *Columbia* accident a misassessment took hold that resisted revision (that is, the misassessment that foam strikes pose only a maintenance problem and not a risk to orbiter safety). It is not simply that the assessment was wrong; what is troubling is the inability to re-evaluate the assessment and re-examine evidence about the vulnerability.

The absence of failure was taken as positive indication that hazards are not present or that countermeasures are effective. In this context, it is very difficult to gather or see if evidence is building up that should trigger a re-evaluation and revision of the organization's model of vulnerabilities. If an organization is not able to change its model of itself unless and until completely clear-cut evidence accumulates, that organization will tend to learn late, that is, it will revise its model of vulnerabilities only after serious events occur. On the other hand, high-reliability organizations assume their model of risks and countermeasures is fragile and even seek out evidence about the need to revise and update this model (Rochlin, 1999). They do not assume their model is correct and then wait for evidence of risk to come to their attention, for to do so will guarantee an organization that acts more riskily than it desires.

The missed opportunities to revise and update the organization's model of the riskiness of foam events seem to be consistent with what has been found in other cases of failure of foresight. We can describe this discounting of evidence as "distancing through differencing," whereby those reviewing new evidence or incidents focus on differences, real and imagined, between the place, people, organization, and circumstances where an incident happens and their own context. By focusing on the differences, people see no lessons for their own operation and practices (or only extremely narrow, well-bounded responses). This contrasts with what has been noted about more effective safety organizations which proactively seek out evidence to revise and update this model, despite the fact that this risks exposing the organization's blemishes.

The distancing through differencing that occurred throughout the build-up to the final Columbia mission can be repeated in the future as organizations and groups look at the analysis and lessons from this accident and the CAIB report. Others in the future can easily look at the CAIB conclusions and deny their relevance to their situation by emphasizing differences (e.g., my technical topic is different, my managers are different, we are more dedicated and careful about safety, we have already addressed that specific deficiency). This is one reason avoiding hindsight bias is so important – when one starts with the question, "How could they have missed what is now obvious?" – one is enabling future distancing through differencing rationalizations. The distancing through differencing process that contributes to this breakdown also indicates ways to change the organization to promote learning. One general principle which could be put into action is – do not discard other events because they appear on the surface to be dissimilar. At some level of analysis all events are unique, while at other levels of analysis they reveal common patterns. Every event, no matter how dissimilar to others on the surface, contains information about underlying general patterns that help create foresight about potential risks before failure or harm occurs. To focus on common patterns rather than surface differences requires shifting the analysis of cases from surface characteristics to deeper patterns and more abstract dimensions. Each kind of contributor to an event can then guide the search for similarities.

This suggests that organizations need a mechanism to generate new evaluations that question the organization's own model of the risks it faces and the countermeasures deployed. Such review and reassessment can help the organization find places where it has underestimated the potential for trouble and revise its approach to create safety. A quasi-independent group is needed to do this – independent enough to question the normal organizational decision-making but involved enough to have a finger on the pulse of the organization (keeping statistics from afar is not enough to accomplish this).

Another general pattern identified in *Columbia* is a fragmented problem-solving process that clouds the big picture. During *Columbia* there was a fragmented view of what was known about the strike and its potential implications. There was no place or person who had a complete and coherent view of the analysis of the foam-strike event including the gaps and uncertainties in the data or analysis to that point. It is striking that people used what looked like technical analyses to justify previously reached conclusions, instead of using technical analyses to test tentative hypotheses.

Discontinuities and internal handovers of tasks increase risk of fragmented problem-solving (Patterson, Roth, Woods, Chow, and Gomez, 2004). With information incomplete, disjointed and patchy, nobody may be able to recognize the gradual erosion of safety constraints on the design and operation of the original system. High reliability organization researchers have found that the importance of free-flowing information cannot be overestimated. A spontaneous and continuous exchange of information relevant to normal functioning of the system offers a background from which signs of trouble can be spotted by those with the experience to do so (Weick, 1993; Rochlin, 1999). Research done on handovers, which is one coordinative device to avert the fragmentation of problem-solving (Patterson, Roth, Woods, Chow, and Gomez, 2004) has identified some of the potential costs of failing to be told, forgetting or misunderstanding information communicated. These costs, for the incoming crew, include:

o   having an incomplete model of the system's state;
o   being unaware of significant data or events;
o   being unprepared to deal with impacts from previous events;
o   failing to anticipate future events;
o   lacking knowledge that is necessary to perform tasks safely;
o   dropping or reworking activities that are in progress or that the team has agreed to do;
o   creating an unwarranted shift in goals, decisions, priorities or plans.

Such problems could also have played a role in the Helios accident, described above. In *Columbia*, the breakdown or absence of cross-checks between disjointed departments and functions is also striking. Cross-checks on the rationale for decisions is a critical part of good organizational decision-making. Yet no cross-checks were in place to detect, question, or challenge the specific flaws in the rationale, and no one noted that cross-checks were missing. The breakdown in basic engineering judgment stands out as well. In *Columbia* the initial evidence available already placed the situation outside the boundary conditions of engineering data and analysis. The only available analysis tool was not designed to predict under these conditions, the strike event was hundreds of times the scale of what the model is designed to handle, and the uncertainty bounds were very large with limited ability to reduce the uncertainty (CAIB, 2003). Being outside the analyzed boundaries should not be confused with not being confident enough to provide definitive answers. In this situation basic engineering judgment calls for large efforts to extend analyses, find new sources of expertise, and cross-check results as Mission Control both practices and does. Seasoned pilots and ship commanders well understand the need for this ability to capture the big picture and not to get lost in a series of details. The issue is how to train for this judgment. For example, the flight director and his or her team practice identifying and handling anomalies through simulated situations. Note that shrinking budgets led to pressure to reduce training investment (the amount of practice, the quality of the simulated situations, and the number or variety of people who go through the simulations sessions can all decline).

What about making technical judgments? Relevant decision-makers did not seem able to notice when they needed more expertise, data, and analysis in order to have a proper evaluation of an issue. NASA's evaluation prior to STS-107 that foam debris strikes do not pose risks of damage to the orbiter demands a technical base. Instead their "resolution" was based on very shaky or absent technical grounds, often with shallow, offhand assessments posing as and substituting for careful analysis.

The fragmentation of problem-solving also illustrates Weick's points about how effective organizations exhibit a "deference to expertise," "reluctance to simplify interpretations," and "preoccupation with potential for failure," none of which was in operation in NASA's organizational decision-making leading up to and during *Columbia* (Weick et al., 1999). A safety organization must ensure that adequate technical grounds are established and used in organizational decision-making. To accomplish this, in part, the safety organization will need to define the kinds of anomalies to be practiced as well as who should participate in simulation training sessions. The value of such training depends critically on designing a diverse set of anomalous scenarios with detailed attention to how they unfold. By monitoring performance in these simulated training cases, safety personnel will be better able to assess the quality of decision-making across levels in the organization.

The fourth pattern in *Columbia* is a failure to revise assessments as new evidence accumulates. The accident shows how difficult it is to revise a misassessment or to revise a once plausible assessment as new evidence comes in. This finding has been reinforced in other studies in different settings (Feltovich et al., 1997; Johnson et al., 1991). Research consistently shows that revising assessments successfully requires a new way of looking at previous facts. Organizations can provide this "fresh" view:

o by bringing in people new to the situation;
o through interactions across diverse groups with diverse knowledge and tools;
o through new visualizations which capture the big picture and reorganize data into different perspectives.

One constructive action is to develop the collaborative interchanges that generate fresh points of view or that produce challenges to basic assumptions. This cross-checking process is an important part of how NASA Mission Control and other organizations successfully respond to anomalies (for a case where these processes break down see Patterson et al., 2004). One can also capture and display indicators of safety margin to help people see when circumstances or organizational decisions are pushing the system closer to the edge of the safety envelope. This idea is something that Jens Rasmussen, one of the pioneers of the new results on error and organizations, has been promoting for two decades (Rasmussen, 1997).

The crux is to notice the information that changes past models of risk and calls into question the effectiveness of previous risk reduction actions, without having to wait for completely clear-cut evidence. If revision only occurs when evidence is overwhelming, there is a grave risk of an organization acting too riskily and finding out only from near-misses, serious incidents, or even actual harm. Instead, the practice of revising assessments of risk needs to be an ongoing process. In this process of continuing re-evaluation, the working assumption is that risks are changing or evidence of risks has been missed.

What is particularly interesting about NASA's organizational decision-making is that the correct diagnosis of production/safety tradeoffs and useful recommendations for organizational change were noted in 2000. The Mars Climate Orbiter report of March 13, 2000, depicts how the pressure for production and to be "better" on several dimensions led to management accepting riskier and riskier decisions. This report recommended many organizational changes similar to those in the CAIB report. A slow and weak response to the previous independent board report was a missed opportunity to improve organizational decision-making in NASA. The lessons of *Columbia* should lead organizations of the future to develop a safety organization that provides "fresh" views on risks to help discover the parent organization's own blind spots and question its conventional assumptions about safety risks.

Finally, the *Columbia* accident brings to the fore another pattern – breakdowns at the boundaries of organizational units. The CAIB analysis notes how a kind of Catch-22 was operating in which the people charged to analyze the anomaly were unable to generate any definitive traction and in which the management was trapped in a stance shaped by production pressure that views such events as turnaround issues. This effect of an "anomaly in limbo" seems to emerge at the boundaries of different organizations that do not have mechanisms for constructive interplay. It is here that we see the operation of the generalization that in risky judgments we have to defer to those with technical expertise and the necessity to set up a problem-solving process that engages those practiced at recognizing anomalies in the event.

This pattern points to the need for mechanisms that create effective overlap across different organizational units and the need to avoid simply staying inside the chain-of-command mentality (though such overlap can be seen as inefficient when the organization is under severe cost pressure). This issue is of particular concern to many organizations as communication technology has linked together disparate groups as a distributed team. This capability for connectivity is leading many to work on how to support effective coordination across these distributed groups, for example in military command and control. A safety organization must have the technical expertise and authority to enhance coordination across the normal chain of command.

ENGINEERING RESILIENCE IN ORGANIZATIONS

The insights derived from the above five patterns and other research results on safety in complex systems point to the need to monitor and manage risk continuously throughout the life-cycle of a system, and in particular to find ways of maintaining a balance between safety and the often considerable pressures to meet production and efficiency goals (Reason,

1997; Weick et al., 1999). These results indicate that safety management in complex systems should focus on resilience – in the face of potential disturbances, changes and surprises, the system's ability to anticipate (knowing what to expect), ability to address the critical (knowing what to look for), ability to respond (knowing what to do), and ability to learn (knowing what can happen). A system's resilience captures the result that failures are breakdowns in the normal adaptive processes necessary to cope with the complexity of the real world (Rasmussen, 1990; Sutcliffe and Vogus, 2003; Hollnagel, Woods, and Leveson, 2006).

A system's resilience includes properties such as:

- buffering capacity: the size or kinds of disruptions the system can absorb or adapt to without a fundamental breakdown in performance or in the system's structure;
- flexibility: the system's ability to restructure itself in response to external changes or pressures;
- margin: how closely the system is currently operating relative to one or another kind of performance boundary;
- tolerance: whether the system gracefully degrades as stress/pressure increase, or collapses quickly when pressure exceeds adaptive capacity.

Cross-scale interactions are another important factor, as the resilience of a system defined at one scale depends on influences from scales above and below: downward in terms of how organizational context creates pressures/goal conflicts/dilemmas and upward in terms of how adaptations by local actors in the form of workarounds or innovative tactics reverberate and influence more strategic issues. Managing resilience, or resilience engineering, then, focuses on what sustains or erodes the adaptive capacities of human-technical systems in a changing environment (Hollnagel et al., 2006). The focus is on monitoring organizational decision-making to assess the risk that the organization is operating nearer to safety boundaries than it realizes (or, more generally, that the organization's adaptive capacity is degrading or lower than the adaptive demands of its environment).

Resilience engineering seeks to develop engineering and management practices to measure sources of resilience, provide decision support for balancing production/safety tradeoffs, and create feedback loops that enhance the organization's ability to monitor/revise risk models and to target safety investments. For example, resilience engineering would monitor evidence that effective cross-checks are well integrated when risky decisions are made, or would serve as a check on how well the organization prepares to handle anomalies by checking on how it practices handling of simulated anomalies (what kind of anomalies, who is involved in making decisions). The focus on system resilience emphasizes the need for proactive measures in safety management: tools to support agile, targeted, and timely investments to defuse emerging vulnerabilities and sources of risk before harm occurs.

To achieve resilience, organizations need support for decisions about production/safety tradeoffs. Resilience engineering should help organizations decide when to relax production pressure to reduce risk, or, in other words, develop tools to support sacrifice decisions across production/safety tradeoffs. When operating under production and efficiency pressures, evidence of increased risk on safety may be missed or discounted. As a result, organizations act in ways that are riskier than they realize or want, until an accident or failure occurs. This is one of the factors that creates the drift toward failure signature in complex system breakdowns.

To make risk a proactive part of management decision-making means knowing when to relax the pressure on throughput and efficiency goals, that is, make a sacrifice decision; how to help organizations decide when to relax production pressure to reduce risk. These tradeoff decisions can be referred to as sacrifice judgments because acute production- or efficiency-related goals are temporarily sacrificed, or the pressure to achieve these goals is relaxed, in order to reduce risks of approaching too near to safety boundary conditions. Sacrifice judgments occur in many settings: when to convert from laparoscopic surgery to an open procedure (e.g., Cook et al., 1998; Woods, 2006), when to break off an approach to an airport during weather that increases the risk of wind shear, or when to have a local slowdown in production operations to avoid risks as complications build up. Ironically, it is at the very times of higher organizational tempo and focus on acute goals that we require extra investment in sources of resilience to keep production/safety tradeoffs in balance – valuing thoroughness despite the potential for sacrifices on efficiency required to meet stakeholder demands.

CONCLUSION

The various models that try to understand safety and "human error" always are works in progress, and their language evolves constantly to accommodate new empirical results, new methods, and new concepts. It is now become obvious, though, that traditional, reductive engineering notions of reliability (that safety can be maintained by keeping system component performance inside acceptable and pre-specified bandwidths) have very little to do with what makes complex systems highly resilient. "Human error" as a label that would indicate a lack of such traditional reliability on part of human components in a complex system, has no analytical leverage whatsoever. Through the various generations of models, "human error" has evolved from cause, to effect, to a mere attribution, that has more to do with those who struggle with a failure in hindsight than with the people caught up in a failing system at the time.

Over the past two decades, research has begun to show how organizations can manage acute pressures of performance and production in a constantly dynamic balance with chronic concern for safety. Safety is not something that these organizations have, it is something that organizations do. Practitioners and organizations, as adaptive systems, continually assess and revise their work so as to remain sensitive to the possibility of failure. Efforts to create safety are ongoing, but not always successful. An organization usually is unable to change its model of itself unless and until overwhelming evidence accumulates that demands revising the model. This is a guarantee that the organization will tend to learn late, that is, revise its model of risk only after serious events occur. The crux is to notice the information that changes past models of risk and calls into question the effectiveness of previous risk reduction actions, without having to wait for complete clear cut evidence. If revision only occurs when evidence is overwhelming, organization will act too riskily and experience shocks from near misses, serious incidents, or even actual harm. The practice of revising assessments of risk needs to be continuous.

Resilience Engineering, the latest addition to thinking about safety and human performance in complex organization, is built on insights derived in part from HRO work, control theory, Perrowian complexity and even man-made disaster theory. It is concerned with assessing organizational risk, that is the risk that organizational decision making will produce unrecognized drift toward failure boundaries. While assessing technical hazards is one kind of input into Resilience Engineering, the goal is to monitor organizational decision making. For example, Resilience Engineering would monitor evidence that effective cross checks are well-integrated when risky decisions are made or would serve as a check on how well the organization is practicing the handling of simulated anomalies (what kind of anomalies, who is involved in making decisions).

Other dimensions of organizational risk include the commitment of the management to balance the acute pressures of production with the chronic pressures of protection. Their willingness to invest in safety and to allocate resources to safety improvement in a timely, proactive manner, despite pressures on production and efficiency, are key factors in ensuring a resilient organization. The degree to which the reporting of safety concerns and problems is truly open and encouraged provides another significant source of resilience within the organization. Assessing the organization's response to incidents indicates if there is a learning culture or a culture of denial. Other dimensions include:

- Preparedness/Anticipation: is the organization proactive in picking up on evidence of developing problems versus only reacting after problems become significant?
- Opacity/Observability – does the organization monitor safety boundaries and recognize how close it is to 'the edge' in terms of degraded defenses and barriers? To what extent is information about safety concerns widely distributed throughout the organization at all levels versus closely held by a few individuals?
- Flexibility/Stiffness – how does the organization adapt to change, disruptions, and opportunities?

Successful organizations in the future will have become skilled at the three basics of Resilience Engineering:

1. detecting signs of increasing organizational risk, especially when production pressures are intense or increasing;
2. having the resources and authority to make extra investments in safety at precisely these times when it appears least affordable;
3. having a means to recognize when and where to make targeted investments to control rising signs of organizational risk and re-balance the safety and production tradeoff.

These mechanisms may help produce an organization that creates foresight about changing risks before failures occur.

PART III
OPERATING AT THE SHARP END

Accidents inevitably lead to close examination of operator actions as post-accident reviewers seek to understand how the accident occurred and how it might have been prevented, especially in high risk settings (e.g., aviation, anesthesia). Practitioner cognition is the source of practitioner actions and so adequate investigation requires teasing apart the interwoven threads of the activation of knowledge relevant to the situation at hand, the flow attention across the multiple issues that arise, and the structure of decisions in the moments leading up to the accident. Accordingly, models of cognition are the tools used to decompose the performance of practitioners into meaningful parts that begin to reveal Second Stories.

There are, to be sure, other motives for developing cognitive models, and there are many available to choose from. What we describe here is a simple framework for decomposing cognitive activities, a guide more than a model, useful in understanding practitioner cognition in semantically complex, time pressured, high consequence domains. Knowledgeable readers will recognize connections to other frameworks for decomposing cognitive activities "in the wild." We introduced this framework as a way to understand expertise and failure in Cook and Woods (1994). We, like almost all in Cognitive Engineering, began with Neisser's (1976) perceptual cycle. Other frameworks in Cognitive Engineering include Rasmussen's classic Skills-Rules-Knowledge (Rasmussen, 1986) and Klein's Recognition Primed Decision Making (Klein et al., 1993). Hutchins (1995a) coined the phrase 'cognition in the wild' to refer to the cognitive activities of people embedded in actual fields of practice.

Knowledgeable readers will also find parts of the simple model proposed here overly crude. The framework does not contain specific, distinct models of internal cognitive mechanisms. Rather it provides a guide to cognitive functions that any cognitive system must perform to handle the demands of complex fields of practice, whether that cognitive system consists of a single individual, a team of people, a system of people and machine agents, or a distributed set of people and machines that communicate and coordinate activities. Developing descriptive models of cognitive functions in context is a basic activity in Cognitive Systems Engineering (see Woods and Roth, 1995 and 1988 for the general case; Roth, Woods and Pople, 1992 for a specific example). Simplicity is a virtue, not just for communicating with non-specialists, in keeping with Hollnagel's Minimal Modeling Manifesto. Throughout the discussions in this book, it is essential to remember that what prompts us to use this particular framework is the need to examine human practitioner performance under stressful, real world conditions to reveal what contributes to success and, sometimes, to failure.

What drives the development of such frameworks for decomposing cognitive activities in context is the requirement to understand accidents. Social need drives the post-accident search for explanations for events with high consequences. Even with cursory examination, it is clear that a variety of factors can play important roles in accident sequences. These factors do not occur at just one level but rather span the full range of human activity. Consider what appears to be a 'simple' case of an anesthesiologist performing an intravenous injection of the 'wrong' drug taken off the backstand holding several different syringes. In this case the practitioner intended to inject a drug used to reverse the effects of neuromuscular blockade but actually injected more of the neuromuscular blocking drug.

The analysis of this incident might proceed along several different paths. The visible salience of a particular cue in the environment, for example, distinctiveness of drug labels, can play a role. However, the structure of multiple tasks and goals can play a role – the practitioner needing to finish this subtask to be able to move on to other goals and their associated tasks. Knowledge is another important factor, especially in semantically complex domains, such as medical practice. Detailed knowledge about how the side effects of one drug can be offset by the primary effects of another can be crucial to successful action sequences. In incidents like this one, three broad categories of cognitive activities help structure the factors that influence practitioner performance: the activation of knowledge, the flow of attention or mindset, and interactions among multiple goals.

It is essential to keep in mind that the point is not to find a single factor which, if different, would have kept the accident from occurring. There are always many such factors. The point is, rather, to develop an exploration of the accident sequence that first captures these different factors and then allows us to see how they play off, one against the other. Thus the purpose of the model is to allow (1) an analysis that structures the reports of accidents (or, more generally, of field observations of operators) into cognitive components and (2) to permit a synthesis that allows us to understand the behavior that results from this cognition.

COGNITIVE SYSTEM FACTORS

The framework used here breaks human cognition in context into three categories. These are knowledge factors, attentional dynamics, and goal conflicts. The next chapters examine each of these in turn, using cases taken from a study of incidents in anesthesia practice. Practitioners will point out that a narrow, isolated view of individual cognition fails to capture the important factors related to work in teams, the use of cognitive tools to produce distributed cognition, and the impact of blunt end factors that shape the world inhabited by sharp end workers. This is surely true; no discussion of individual human cognition will be adequate for an understanding of the way that practitioners accomplish their work in the real world. But any discussion of human cognition must start somewhere and the nature of actual practice inexorably leads to broader systems issues. The framework used here is focused on giving some coherence to the many isolated observations about how people solve problems, cope with failure or impending failure, structure their cognitive and physical environments.

6

BRINGING KNOWLEDGE TO BEAR IN CONTEXT

CONTENT, ORGANIZATION AND ACTIVATION OF KNOWLEDGE

As the name suggests, this chapter deals with practitioners' knowledge of the technical system. This part of knowledge factors is more or less familiar ground for decision theorists, expert system designers, and technical experts. But in the arena of cognitive analysis, knowledge factors is a broader category. It also includes the organization and application of the knowledge, that is, the extent to which the knowledge can be used flexibly in different contexts. It includes an examination of the types of processes that "call to mind" specific items of knowledge relevant to the situation at hand. In other words, the category encompasses the way that practitioners bring knowledge to bear effectively in their cognitive work. This work includes decision making and problem solving, but also what are usually considered the relatively mundane aspects of daily practice that involve application of knowledge.

> ### CASE 6.1   MYOCARDIAL INFARCTION
>
> An elderly patient presented with impending limb loss, specifically a painful, pulseless, blue arm indicating an arterial thrombus blood clot in one of the major arteries that threatened loss of that limb. The medical and surgical history were complicated and included hypertension, insulin dependent diabetes mellitus, a myocardial infarction and prior coronary artery bypass surgery. There was clinical and laboratory evidence of worsening congestive heart failure: shortness of breath, dyspnea on exertion and pedal edema. Electrocardiogram (ECG) changes included inverted T waves. In the emergency room a chest x-ray suggested pulmonary edema, the arterial blood gas (ABG) showed markedly low oxygen tension (PaO$_2$ of 56 on unknown FiO$_2$), and the blood glucose was 800. The patient received furosemide (a diuretic) and 12 units of regular

insulin in the emergency room. There was high urine output.

The patient was taken to the operating room for removal of the clot under local anesthesia with sedation provided by the anesthetist. In the operating room the patient's blood pressure was high, 210/120; a nitroglycerine drip was started and increased in an effort to reduce the blood pressure. The arterial oxygen saturation (SaO_2) was 88% on nasal cannula and did not improve with a rebreathing mask, but rose to the high 90s when the anesthesia machine circuit was used to supply 100% oxygen by mask. The patient did not complain of chest pain but did complain of epigastric pain and received morphine. Urine output was high in the operating room. The blood pressure continued about 200/100. Nifedipine was given sublingually and the pressure fell over 10 minutes to 90 systolic. The nitroglycerine was decreased and the pressure rose to 140. The embolectomy was successful. Postoperative cardiac enzyme studies showed a peak about 12 hours after the surgical procedure indicating that the patient had suffered a heart attack sometime in the period including the time in the emergency room and the operating room. The patient survived.

The patient, a person with known heart disease, prior heart attack and heart surgery, required an operation to remove a blood clot from the arm. There were signs of congestive heart failure, for which he was treated with a diuretic, and also of out-of-control blood sugar, for which he was treated with insulin, and cardiac angina. In an effort to get the blood pressure under control, the patient was first given one drug and then another, stronger medicine. The result of this stronger medicine caused severe low blood pressure. There was later laboratory evidence that the patient had another heart attack sometime around the time of surgery.

In this incident, the anesthetist confronted several different conditions. The patient was acutely ill, dangerously so, and would not be a candidate for an elective surgical procedure. A blood clot in the arm, however, was an emergency: failing to remove it will likely result in the patient losing the arm. There were several problems occurring simultaneously. The arterial blood gas showed markedly low oxygen. Low oxygen in the blood meant poor oxygen delivery to the heart and other organs. High blood pressure created a high mechanical workload for the heart. But low blood pressure was also undesirable because it would reduce the pressure to the vessels supplying the heart with blood (which, in turn, supplies the heart with oxygen).

To deal with each of these issues the practitioner was employing a great deal of knowledge (in fact, the description of just a few of the relevant aspects of domain knowledge important to the incident would occupy several pages). The individual actions of the practitioner can each be traced to specific knowledge about how various physiological and pharmacological systems work; the actions are grounded in knowledge. The question for us in this case is how the knowledge is organized and how effectively it is brought to bear.

Significantly, the issues in this case are not separate but interact in several ways important to the overall state of the patient. Briefly, the high glucose value indicated diabetes out of control; when the blood sugar is high, there is increased urine output as the glucose draws water into the urine. The diuretic given in the emergency room added to the creation of urine. Together, these effects create a situation in which the patient's intravascular volume (the amount of fluid in the circulatory system) was low. The already damaged heart (prior heart attack) is also starved for oxygen (low arterial oxygen tension). The patient's pain leads to high blood pressure, increasing the strain on the heart.

There is some evidence that the practitioner was missing or misunderstanding important features of the evolving situation. It seems (and seemed to peer experts who evaluated the incident at the time; cf., Cook et al., 1991) that the practitioner misunderstood the nature of the patient's intravascular volume, believing the volume was high rather than low. The presence of high urine output, the previous use of a diuretic (furosemide, trade name Lasix) in the emergency room, and the high serum glucose together are indications that a patient should be treated differently than was the case here. The high glucose levels indicated a separate problem that seemed to be unappreciated by the practitioner on the scene. In retrospect, other practitioners argued that the patient probably should have received more intravenous fluid and should have been monitored using more invasive monitoring to determine when enough fluid had been given (e.g., via a catheter that goes through the heart and into the pulmonary artery).

But it is also apparent that many of the practitioner's actions were appropriate in the context of the case as it evolved. For example, the level of oxygen in the blood was low and the anesthetist pursued several different means of increasing the blood oxygen level. Similarly, the blood pressure was high and this, too, was treated, first with nitroglycerin (which may lower the blood pressure but also can protect the heart by increasing its blood flow) and then with nifedipine. The fact that the blood pressure fell much further than intended was probably the result of depleted intravascular volume which was, in turn, the result of the high urinary output provoked by the diuretic and the high serum glucose level. Significantly, the treatment in the emergency room that preceded the operation made the situation worse rather than better.

In the opinion of anesthesiologist reviewers of this incident shortly after it occurred, the circumstances of this case should have brought to mind a series of questions about the nature of the patient's intravascular volume. Those questions would then have prompted the use of particular monitoring techniques before and during the surgical procedure.

This incident raises a host of issues regarding how knowledge factors affect the expression of expertise and error. Bringing knowledge to bear effectively in problem solving is a process that involves:

○ content (what knowledge) – is the right knowledge there? is it incomplete or erroneous (i.e., "buggy");
○ organization – how knowledge is organized so that relevant knowledge can be activated and used effectively; and

○   activation – is relevant knowledge "called to mind" in different contexts.

Much attention is lavished on content but, as this incident demonstrates, mere possession of knowledge is not expertise. Expertise involves knowledge organization and activation of knowledge in different contexts (Bransford, Sherwood, Vye, and Rieser, 1986). Moreover, it should be clear from the example that the applications of knowledge go beyond simply matching bits of knowledge to specific items in the environment. The exact circumstances of the incident were novel, in the sense that the practitioner had never seen precisely this combination of conditions together in a single patient, but we understand that human expertise involves the flexible application of knowledge not only for familiar, repetitive circumstances but also in new situations (Feltovich, Spiro and Coulson, 1989).

When analyzing the role of knowledge factors in practitioner performance, there are overlapping categories of research that can be applied. These include:

○   mental models and knowledge flaws (sometimes called "buggy" knowledge),
○   knowledge calibration,
○   inert knowledge, and
○   heuristics, simplifications, and approximations.

MENTAL MODELS AND "BUGGY" KNOWLEDGE

Knowledge of the world and its operation may be complete or incomplete. It may also be accurate or inaccurate. Practitioners can only act on the knowledge they have. The notion of a mental model, that is a mental representation of the way that the (relevant part of the) world works is now well established, even if researchers do not agree on how such a model is developed or maintained. What is clear, however, is that the function of such models is to order the knowledge of the work so as to allow the practitioner to make useful inferences about what is happening, what will happen next, and what *can* happen. The term mental model is particularly attractive because it acknowledges that things in the world are related, connected together in ways that interact, and that it is these interactions that are significant, rather than some isolated item of knowledge, discrete and separate from all others.

Of course the mental model a practitioner holds may be incomplete or inaccurate. Indeed, it is clear that all such models are imperfect in some ways – imprecise or, more likely, incomplete. Moreover, mental models must contain information that is nowhere in textbooks but learned and refined through experience. How long it takes the sublingual nifedipine to work, in this incident, and how to make inferences back from the occurrence of low blood pressure to the administration of the nifedipine earlier is an example. When practitioner mental models are inaccurate or incomplete they are described as "buggy" (see Gentner and Stevens, 1983; Rouse and Morris, 1986; Chi, Glaser, and Farr, 1988, for some of the basic results on mental models).

For example, Sarter and Woods (1992, 1993) found that buggy mental models contributed to problems with cockpit automation. A detailed understanding of the various modes of flight deck automation is a demanding knowledge requirement for pilots in highly automated cockpits. Buggy mental models played a role in automation surprises, cases where pilots were "surprised" by the automation's behavior. The buggy knowledge created flaws in the understanding the automatic system's behavior. Buggy mental models made it hard for pilots to determine what the automation was doing, why it was doing it, and what it would do next. Nearly the same problems are found in other domains, such as anesthesiologists using microcomputer-based devices (Cook, Potter, Woods, and McDonald, 1991b).

Significantly, once the possibility of buggy mental models is recognized, it is possible to design experiments that reveal specific bugs or gaps. By forcing pilots to deal with various non-normal situations in simulator studies, it was possible to reveal knowledge bugs and their consequences. It is also possible to find practitioners being sensitive to these gaps or flaws in their understanding and adapting their work routines to accommodate these flaws. In general, when people use "tried and true" methods and avoid "fancy features" of automation, we suspect that they have gaps in their models of the technology. Pilots, for example, tend to adopt and stay with a small repertoire of strategies, in part, because their knowledge about the advantages and disadvantages of the various options for different flight contexts is incomplete. But these strategies are themselves limiting. People maybe aware of flaws in their mental models and seek to avoid working in ways that will give those flaws critical importance, but unusual or novel situations may force them into these areas.

It is not clear in the incident described whether the practitioner was aware of the limitations of his mental model. He certainly did not behave as though he recognized the consequences of the interdependent facets of the problem.

Technology Change and Knowledge Factors

All of the arenas where human error is important intensively use technology. Significantly, the technological strata on which such domains are based are more or less in constant flux. In medicine, transportation and other areas, technological change is constant. This change can have important impacts on knowledge factors in a cognitive system. First, technology change can introduce substantial new knowledge requirements. For example, pilots must learn and remember the available options in new flight computers, learn and remember how to deploy them across a variety of operational circumstances – especially during the rare but difficult or critical situations, learn and remember the interface manipulations required to invoke the different modes or features, learn and remember how to interpret or where to find the various indications about which option is active or armed and the associated target values entered for each. And here by the word "remember" we mean not simply being able to demonstrate the knowledge in some formal way but rather be able to call it to mind and use it effectively in actual task contexts.

Studying practitioner interaction with devices is one method for understanding how people develop, maintain, and correct flaws in mental models. Because so much of practitioner action in mediated through devices (e.g., cockpit controls) flaws in mental models here tend to have severe consequences.

Several features of practitioner interactions with devices suggest more general activities in the acquisition and maintenance of knowledge.

1. *Knowledge extension by analogy*: Users transfer their mental models developed to understand past devices to present ones if the devices appear to be similar, even if the devices are internally dissimilar.
2. *There is no cognitive vacuum*: Users' mental models are based on inferences derived from experience with the apparent behavior of the device, but these inferences may be flawed. Devices that are "opaque" and that give no hint about their structure and function will still be envisioned as having some internal mechanism, even if this is inaccurate. Flaws in the human-computer interface may obscure important states or events or incidentally create the appearance of linkages between events or states that are not in fact linked. These will contribute to buggy mental models of device function.
3. *Each experience is an experiment*: Practitioners use experience with devices to revise their models of device operations. They may do this actively, by deliberately experimenting with ways of using the device or passively by following the behavior of the device over time and making inferences about its function. People are particularly sensitive to apparent departures from what "normal".
4. *Hidden complexity is treated as simplicity*: Devices that are internally complex but superficially simple encourage practitioners to adopt overly simplistic models of device operation and to develop high confidence that these models are accurate and reliable.

Device knowledge is a large and readily identified area where knowledge defects can be detected and described. But characteristics of practitioner interaction with devices have parallels in the larger domain. Thus, the kinds of behaviors observed with devices are also observed in use of knowledge more generally.

Knowledge Calibration

Closely related to the last point above are results from several studies (Sarter and Woods, 1993; Cook et al., 1991; Moll van Charante et al., 1993) indicating that practitioners are often unaware of gaps or bugs in their mental models. This lack of awareness of flaws in knowledge broadly the issue of *knowledge calibration* (e.g., Wagenaar and Keren, 1986).

Put most simply, individuals are well calibrated if they are aware of the accuracy, completeness, limits, and boundaries of their knowledge, i.e., how well they know what they know. People are miscalibrated if they are overconfident (or much less commonly underconfident) about the accuracy and compass of their knowledge. Note that degree of calibration is not the same thing as expertise; people can be experts in part because they are well calibrated about where their knowledge is robust and where it is not.

There are several factors that can contribute to miscalibration. First, the complexity of practice means that areas of incomplete or buggy knowledge can remain hidden from practitioners for long periods. Practitioners develop habitual patterns of activity that become well practiced and are well understood. But practitioners may be unaware that their knowledge outside these frequently used regions is severely flawed simply because they never have occasion to need this knowledge and so never have experience with its inaccuracy or limitations. Practitioners may be able to arrange their work so that situations which challenge their mental models or confront their knowledge are limited. Second, studies of calibration indicate that the availability of feedback, the form of feedback and the attentional demands of processing feedback, can affect knowledge calibration (e.g., Wagenaar and Keren, 1986). Even though flaws in practitioner knowledge are being made apparent by the failure, so much attention may be directed to coping with failure that the practitioner is unable to recognize that his or her knowledge is buggy and so recalibration never occurs.

Problems with knowledge calibration, rather than simply with lack of knowledge, may pose substantial operational hazards. Poor calibration is subtle and difficult for individuals to detect because they are, by definition, unaware that it exists.

Avoiding miscalibration requires that information about the nature of the bugs and gaps in mental models be made apparent through feedback. Conversely, systems where feedback is poor have a high propensity for maintaining miscalibrated practitioners. A relationship between poor feedback and miscalibrated practitioners was found in studies of pilot-automation interaction (Sarter and Woods, 1993) and of physician-automation interaction (Cook and Woods, 1996b). For example, some of the participants in the former study made comments in the post-scenario debriefings such as: "I never knew that I did not know this. I just never thought about this situation." Although this phenomenon is most easily demonstrated when practitioners attempt to use computerized devices because such devices so often are designed with opaque interfaces, it is ubiquitous.

Knowledge miscalibration is especially important in the discussion of error. Failures that occur in part because of miscalibration are likely to be reported as other sorts of failures; the absent knowledge stays absent and unregarded. Thus problems related to, for example, poorly designed devices go unrecognized. Significantly, the ability to adequately reconstruct and examine the sequence of events following accidents is impaired: the necessary knowledge is absent but those involved in the accident are unaware of this absence and will seek explanations from other sources.

Activating Relevant Knowledge in Context: The Problem of Inert Knowledge

A more subtle form of knowledge problem is that of *inert knowledge,* that is knowledge that is not accessed and remains unused in important work contexts. This problem may play a role in incidents where practitioners know the individual pieces of knowledge needed to build a solution but are unable to join the pieces together because they have not confronted the need previously. (Note that inert knowledge is a concept that overlaps both knowledge and attention in that it refers to knowledge that is present in some form but not activated in the appropriate situation. The interaction of the three cognitive factors is the norm.) Thus, the practitioner in the first incident could be said to know about the relationship between blood glucose, furosemide, urine output, and intravascular volume but also not to know about that relationship in the sense that the knowledge was not activated at the time when it would have been useful. The same pattern can occur with computer aids and automation. For example, some pilots were unable to apply knowledge of automation successfully in an actual flight context despite the fact that they clearly possessed the knowledge as demonstrated by debriefing, that is, their knowledge was inert (Sarter and Woods, 1993).

We tend to assume that if a person can be shown to possess a piece of knowledge in one situation and context, then this knowledge should be accessible under all conditions where it might be useful. But there are a variety of factors that affect the activation and use of relevant knowledge in the actual problem solving context (e.g., Bransford et al., 1986). But it is clear that practitioners may experience dissociation effects where the retrieval of knowledge depends on contextual cues (Gentner and Stevens, 1983; Perkins and Martin, 1986). This may well have been the case in the first incident. During later discussion, the practitioner was able to explain the relationship between the urine output, hyperglycemia, diuretic drugs, and intravascular volume and in that sense possessed the relevant knowledge, but this knowledge was not summoned up during the incident.

Results from accident investigations often show that the people involved did not call to mind all the relevant knowledge during the incident although they "knew" and recognized the significance of the knowledge afterwards. The triggering of a knowledge item X may depend on subtle pattern recognition factors that are not present in every case where X is relevant. Alternatively, that triggering may depend critically on having sufficient time to process all the available stimuli in order to extract the pattern. This may explain the difficulty practitioners have in "seeing" the relevant details when the pace of activity is high and there are multiple demands on the practitioner. These circumstances are typical of systems "at the edge of the performance envelope."

The problem of inert knowledge is especially troubling because it is so difficult to determine beforehand all the situations in which specific knowledge needs to be called to mind and employed. Instead, we rely on relatively static recitals of knowledge (e.g., written or oral examinations) as demonstrations of practitioner knowledge. From a cognitive analysis perspective, what is critical is to show that the problem solver can and does access situation-relevant knowledge under the conditions in which tasks are performed.

Oversimplifications

One means for coping with complexity is the use of simplifying heuristics. Heuristics amount to cognitive "rules of thumb", that is approximations or simplifications that are easier to apply than more formal decision rules. Heuristics are useful because they are easy to apply and minimize the cognitive effort required to produce decisions. Whether they produce desirable results depends how well they work, that is, how satisfactorily they allow practitioners to produce good cognitive performance over a variety of problem demand factors (Woods, 1988). In all cases heuristics are to some degree distortions or misconceptions – if they were not, they would not be heuristics but rather robust rules. It is possible for heuristics that appear to work satisfactorily under some conditions to produce "error" in others. Such heuristics amount to "oversimplifications."

In studying the acquisition and representation of complex concepts in biomedicine, Feltovich et al. (1989) found that some medical students (and even by some practicing physicians) applied knowledge to certain problems in ways that amounted to oversimplification. They found that "bits and pieces of knowledge, in themselves sometimes correct, sometimes partly wrong in aspects, or sometimes absent in critical places, interact with each other to create large-scale and robust misconceptions" (Feltovich et al., 1989, p. 162). Broadly, oversimplifications take on several different forms (see Feltovich, Spiro, and Coulson, 1993):

- seeing different entities as more similar than they actually are,
- treating dynamic phenomena statically,
- assuming that some general principle accounts for all of a phenomenon,
- treating multidimensional phenomena as unidimensional or according to a subset of the dimensions,
- treating continuous variables as discrete,
- treating highly interconnected concepts as separable,
- treating the whole as merely the sum of its parts.

Feltovich and his colleagues' work has important implications for the teaching and training. In particular, it challenges what might be called the "building block" view of learning where initially lessons present simplified material in modules that decompose complex concepts into their simpler components with the belief that these will eventually "add up" for the advanced learner (Feltovich et al., 1993). Instructional analogies, while serving to convey certain aspects of a complex phenomenon, may miss some crucial ones and mislead on others. The analytic decomposition misrepresents concepts that have interactions among variables. The conventional approach can produce a false sense of understanding and inhibit pursuit of deeper understanding. Learners resist learning a more complex model once they already have an apparently useful simpler one (Spiro et al., 1988).

But the more basic question associated with oversimplification remains unanswered. Why do practitioners utilize simplified or oversimplified knowledge at all? Why don't practitioners use formal rules based, for example, on Bayesian decision theoretical reasoning? The answer is that the simplifications offered by heuristics reduce the cognitive effort required in demanding circumstances.

> It is easier to think that all instances of the same nominal concept ... are the same or bear considerable similarity. It is easier to represent continuities in terms of components and steps. It is easier to deal with a single principle from which an entire complex phenomenon "spins out" than to deal with numerous, more localized principles and their interactions. (Feltovich et al., 1989, p. 131)

This actually understates the value of heuristics. In some cases, it is apparent that the heuristics produce better decision making over time than the formally "correct" processes of decision making. The effort required to follow more "ideal" reasoning paths may be so large that it would keep practitioners from acting with the speed demanded in actual environments. When the effort required to reach a decision is included and the amount of resource that can be devoted to decision making is limited (as it is in real world settings), heuristics can actually be superior to formal rule following. Payne, Bettman, and Johnson (1988) and Payne, Johnson, Bettman, and Coupey (1990) demonstrated that simplified methods produce a higher proportion of correct choices between multiple alternatives under conditions of time pressure than do formal Bayesian approaches that require calculation. Looking at a single instance of failure may lead us to conclude that the practitioner made an "error" because he or she did not apply an available, robust decision rule. But the error may actually be ours rather than the practitioners when we fail to recognize that using such (effortful) procedures in all cases will actually lead to a greater number of failures than application of the heuristic!

There is a more serious problem with an oversimplified view of oversimplification by practitioners. This is our limited ability to account for uncertainties, imprecision, or conflicts that need to be resolved in individual cases. In the incident, for example, there are conflicts between the need to keep the blood pressure high and the need to keep the blood pressure low. As is often the case in this and similar domains, the locus of conflict may vary from case to case and from moment to moment. The heart depends on blood pressure for its own blood supply, but increasing the blood pressure also increases the work it is required to perform. The practitioner must decide what blood pressure is acceptable. Many factors enter into this decision process. For example, how precisely can we predict the future blood pressure? How will attempts to reduce blood pressure affect other physiological variables? How is the pressure likely to change without therapy? How long will the surgery last? Will changes in the blood pressure impact other systems (e.g., the brain)? Only in the world of the classroom (or courtroom) can such questions be regarded as answered in practice because they can be answered in principle. The complexity of real practice means that virtually all approaches will appear, when viewed from a decision theoretical perspective, to be oversimplifications – it is a practical impossibility before the fact to produce exhaustively complete and robust rules for performance. (The marked failure of computer based decision tools to handle cases such as the incident presented in this section is evidence, if more were needed, about the futility of searching for a sufficiently rich and complicated set of formal rules to define "good" practice.)

In summary, heuristics represent effective and necessary adaptations to the demands of real workplaces (Rasmussen, 1986). When post-incident cognitive analysis points to practitioner oversimplification we need to examine more than the individual incident in order to determine whether decision making was flawed. We cannot simply point to an available formal decision rule and claim that the "error" was the failure to apply this (now apparently) important formal decision rule. The problem is not *per se* that practitioners use shortcuts or simplifications, but that their limitations and deficiencies were not apparent. Cognitive analysis of knowledge factors therefore is extended examination of the ways practitioners recognize situations where specific simplifications are no longer relevant, and when (and how) they know to shift to using more complex concepts, methods, or models.

ANALYZING THE COGNITIVE PERFORMANCE OF PRACTITIONERS FOR KNOWLEDGE FACTORS

The preceding discussion has hinted at the difficulty we face when trying to determine how buggy mental models, oversimplifications, inert knowledge, or some combination contributed to an incident. The kinds of data available about the incident evolution, the knowledge factors for the specific practitioners involved in the incident, the knowledge factors in the practitioner population in general, are critical to our understanding of the human performance in the incident. These sorts of high precision data are rarely available without special effort from investigators and researchers. The combination of factors present in Incident 1 was unusual, and this raises suspicion that a buggy mental model of the relationship between these factors played a major role. But the other characteristic flaws that can occur under the heading of knowledge factors are also likely candidates.

Given the complexities of the case, oversimplification strategies could be implicated. The congestive heart failure is usually associated with increased circulating blood volume and the condition is improved by diuretic therapy. But in this case high blood glucose was already acting as a diuretic and the addition of the diuretic drug furosemide (which occurred in the emergency room before the anesthesia practitioner had contact with the patient) probably created a situation of relative hypovolemia, that is too little rather than too much. The significance of the earlier diuretic in combination with the diabetes was missed, and the practitioner was unable to recognize how this situation varied from typical for congestive heart failure.

Inert knowledge may have played a role as well. The cues in this case were not the ones that are usually associated with deeper knowledge about the inter-relationships of intravascular volume, glucose level, and cardiovascular volume. The need to pay attention to the patient's low oxygen saturation and other abnormal conditions may well have contributed to making some important knowledge inert.

Beyond being critical of practitioner performance from afar, we might ask how the practitioners themselves view this sort of incident. How clearly does our cognitive analysis correspond to their own understanding of human performance. Interestingly, practitioners are acutely aware of how deficient their rules of thumb may be, how susceptible to failure are the simplifications they use to achieve efficient performance. Practitioners are actually aware that certain situations may require abandoning a cognitively less effortful approach in favor of more cognitively demanding "deep thinking." For example, senior anesthesiologists commenting on the first incident shortly after it occurred were critical of practitioner behavior:

> This man was in major sort of hyperglycemia and with popping in extra Lasix [furosemide] you have a risk of hypovolemia from that situation. I don't understand why that was quietly passed over, I mean that was a major emergency in itself ... this is a complete garbage amount of treatment coming in from each side, responding from the gut to each little bit of stuff [but it] adds up to no logic whatsoever ... the thing is that this patient [had] an enormous number of medical problems going on which have been simply reported [but] haven't really been addressed.

This is a pointed remark, made directly to the participant in a large meeting by those with whom he worked each day. While it is not couched in the language of cognitive science, it remains a graphic reminder that practitioners recognize the importance of cognition to their success and sometimes distinguish between expert and inexpert performance by looking for evidence of cognitive processes.

7
MINDSET

ATTENTIONAL DYNAMICS

Mindset is about attention and its control (Woods, 1995b). It is especially critical when examining human performance in dynamic, evolving situations where practitioners are required to shift attention in order to manage work over time. In all real world settings there are multiple signals and tasks competing for practitioner attention. On flight decks, in operating rooms, or shipboard weapon control centers, attention must flow from object to object and topic to topic. Sometimes intrusions into practitioner attention are distractions but other times they are critical cues that important new data is available (Klein, Pliske, et al., 2005). There are a host of issues that arise under this heading. Situation awareness, the redirection of attention amongst multiple threads of ongoing activity, the consequences of attention being too narrow (fixation) or too broad (vagabonding) – all critical to practitioner performance in these sorts of domains – all involve the flow of attention and, more broadly, mindset. Despite its importance, understanding the role of mindset in accidents is difficult because in retrospect and with hindsight investigators know exactly what was of highest priority when.

CASE 7.1 HYPOTENSION

During a coronary artery bypass graft procedure an infusion controller device used to control the flow of a sodium nitroprusside (SNP) to the patient delivered a large volume of drug at a time when no drug should have been flowing. Five of these microprocessor-based devices, each controlling the flow of a different drug, were set up in the usual fashion at the beginning of the day, prior to the beginning of the case. The initial part of the case was unremarkable. Elevated systolic blood pressure (>160 torr) at the time of sternotomy prompted the practitioner to begin an infusion of SNP. After starting the infusion at 10 drops per minute, the device began

> to sound an alarm. The tubing connecting the device to the patient
> was checked and a stopcock (valve) was found closed. The operator
> opened the stopcock and restarted the device. Shortly after restart, the
> device alarmed again. The blood pressure was falling by this time, and
> the operator turned the device off. Over a short period, hypertension
> gave way to hypotension (systolic pressure <60 torr). The hypotension was
> unresponsive to fluid challenge but did respond to repeated injections
> of neosynephrine and epinephrine. The patient was placed on bypass
> rapidly. Later, the container of nitroprusside was found to be empty; a full
> bag of 50 mg in 250 ml was set up before the case.
>
> The physicians involved in the incident were comparatively
> experienced device users. Reconstructing the events after the incident
> led to the conclusion that the device was assembled in a way that would
> allow free flow of drug. Initially, however, the stopcock blocked drug
> delivery. The device was started, but the machine did not detect any
> flow of drug (because the stopcock was closed) and this triggered visual
> and auditory alarms. When the stopcock was opened, free flow of fluid
> containing drug began. The controller was restarted, but the machine
> again detected no drip rate, this time because the flow was a continuous
> stream and no individual drops were being formed. The controller alarmed
> again with the same message that had appeared to indicate the earlier
> no flow condition. Between opening the stopcock and the generation of
> the error message, sufficient drug was delivered to substantially reduce the
> blood pressure. The operator saw the reduced blood pressure, concluded
> that the SNP drip was not required and pushed the control button marked
> "off." This powered down the device, but the flow of drug continued. The
> blood pressure fell even further, prompting a diagnostic search for sources
> of low blood pressure. The SNP controller was seen to be off. Treatment of
> the low blood pressure itself commenced and was successful.

We need to point out that a focus on "error" in this incident would distract our attention from the important issues the incident raises. The human performance during the incident was both flawed and exemplary. The failures in working with the infusion device contrast markedly with the successes in managing the complications. Even though they were unable to diagnose its source, the practitioners were quick to correct the physiologic, systemic threat, shifting their focus of attention. Their ability to shift from diagnosis to disturbance management was crucial to maintaining the system (Woods, 1988; Woods and Hollnagel, 2006, Chapter 8).

Mindset, where attention is focused and how it shifts and flows over time, is critical in this incident. There were many things happening in the world and it was impossible to attend to them all. But clearly the important thing that was happening went unregarded. That is to say attention flowed not to it but to other things.

Why the device did not receive attention? There are numerous contributors to the inability to observe and correct the unintended flow of drug via the infusion device. Some more obvious ones related to the device itself are:

1. the drip chamber was obscured by the machine's sensor, making visual inspection difficult,
2. presence of an aluminum shield around the fluid bag, hiding its decreasing volume,
3. misleading alarm messages from the device,
4. presence of multiple devices making it difficult to trace the tubing pathways.

The practitioners reported that they turned the device off as soon as the blood pressure fell, at about the same moment that the device alarmed a second time. In their mindset, the device was off and unimportant. It was unregarded, unattended to. When we say it was *unimportant* we do not mean that it was examined and found to be uninvolved in the evolving situation. This would have required attention to flow to the device (at least briefly) so that inferences about its state could be made. The post-incident practitioner debriefing leads us to believe that they dismissed the device from their mindset once it was "off" – the device played no role in the flow of their attention from that point. The practitioners did not make inferences about how the device was working or not working or how it might be playing a role: they did not attend to the device at all. Once it was turned "off" it disappeared from practitioner attention, not becoming the focus of attention again until the very end of the sequence. The device was absent from the mindset of the practitioners. This is not to say that the practitioners were idle or inattentive, indeed they engaged in diagnostic search for the causes of low blood pressure and the management of that complication. It is definitely not the case that they were inattentive but rather that their mindset did not include the relevant object and its (mis)function.

It is easy to speculate on alternative sequences that might have occurred if circumstances had been slightly different. We are compelled to address the device design, its placement in the room, the other distractors that are present in such settings, and a host of other "obvious" problems. If the device had been more robustly designed, it might have not had this particular failure mode. If the users had experience with this particular form of device failure they might have retained the device in their mindset as a (potentially) active influence in the situation. If the situation had been less demanding, less fast paced, the meaning of the device "off" state might have been questioned. Here the "off" condition was indicated by a blank liquid crystal diode display screen. The "on-off" state was controlled by a push button that toggled the state from on to off and back, this might have included some uncertainty (e.g., that the device was powered down but that fluid flow was possible). But rather than focus on these issues, our attention should now turn to the issue of mindset, its construction, maintenance, strengths and weaknesses. How is attention directed by experience? How is mindset established? How do practitioners adjust their processes of attending to different stimuli based on their mindset?

The control of attention is an important issue for those trying to understand human performance, especially in event-rich domains such as flightdecks, operating rooms, or control centers. Attention is a limited resource. One cannot attend to more than one thing at a time, and so shifts of attention are necessary to be able to "take in" the ways in which the world is changing. When something in the world is found that is anomalous (what is sensed in the world is not consistent with what is expected by the observer) attention focuses on that thing and a process of investigation begins that involves other shifts of attention. This process is ongoing and has been described by Neisser as the perceptual or cognitive cycle (Neisser, 1976). It is a crucial concept for those trying to understand human performance because it is the basis for all diagnosis and action. Nothing can be discovered in the world without attention; no intended change in the world can be effected without shifting attention to the thing being acted upon. At least two kinds of human performance problems are based on attentional dynamics. The first is a loss of situation awareness and the second is psychological fixation.

"LOSS OF SITUATION AWARENESS"

Situation awareness is a label that is often used to refer to many of the cognitive processes involved in what we have called here attentional dynamics (Endsley, 1995; Sarter and Woods, 1991; Adams, Tenney, and Pew, 1995; Woods and Sarter, 2010). There have been many debates about what is situation awareness and attempts to measure it as a unitary phenomenon. For example, does situation awareness refer to a product or a process? It is not our intention here to engage in or outline a position in these debates. Here we are using the label situation awareness, since it is a commonly used expression, to point to the cognitive processes involved in the control of attention. Just a few of the cognitive processes that may be involved when one invokes the label of situation awareness are: control of attention (Gopher, 1991), mental simulation (Klein and Crandall, 1995), forming expectancies (Johnson, Grazioli, Jamal, and Zualkernan, 1992; Christoffersen, Woods and Blike, 2007), directed attention (Woods, 1995b), and contingency planning (Orasanu, 1990). Because the concept involves tracking processes in time, it can also be described as mental bookkeeping – keeping track of multiple threads of different but interacting sub-problems as well as of influences of the activities undertaken to control them (Cook, Woods and McDonald, 1991; Woods and Hollnagel, 2006).

Maintaining situation awareness necessarily requires shifts of attention between the various threads. It also requires more than attention alone, for the objective of the shifts of attention is to inform and modify a coherent picture or model of the system as a whole. Building and maintaining that picture require cognitive effort. Breakdowns in these cognitive processes can lead to operational difficulties in handling the demands of dynamic, event-driven incidents. In aviation circles this is known as "falling behind the plane" and in aircraft carrier flight operations it has been described as "losing the bubble" (Roberts and Rousseau, 1989). In each case what is being lost is the operator's internal representation of the state of the world at that moment and the direction in which the forces active in the world are taking the system that the operator is trying to control. Dorner (1983) calls breakdowns in mental bookkeeping "thematic vagabonding" as the practitioner jumps from thread to thread in an uncoordinated fashion (the response in Incident #1 may have possessed an element of vagabonding).

Fischer, Orasanu, and Montvalo (1993) examined the juggling of multiple threads of a problem in a simulated aviation scenario. More effective crews were better able to coordinate their activities with multiple issues over time; less effective crews traded one problem for another. More effective crews were sensitive to the interactions between multiple threads involved in the incident; less effective crews tended to simplify the situations they faced and were less sensitive to the constraints of the particular context they faced. Less effective crews "were controlled by the task demands" and did not look ahead or prepare for what would come next. As a result, they were more likely to run out of time or encounter other cascading problems. Interestingly, there were written procedures for each of the problems the crews faced. The cognitive work associated with managing multiple threads of activity is different from the activities needed to merely follow the rules.

Obtaining a clear, empirically testable model for situation awareness is difficult. For example, Hollister (1986) presents an overview of a model of divided attention operations – tasks where attention must be divided across a number of different input channels and where the focus of attention changes as new events signal new priorities. This model then defines an approach to breakdowns in attentional dynamics (what has been called a divided attention theory of error) based on human divided attention capabilities balanced against task demands and adjusted by fatigue and other performance-shaping factors. Situation awareness is clearly most in jeopardy during periods of rapid change and where a confluence of forces makes an already complex situation critically so. This condition is extraordinarily difficult to reproduce convincingly in a laboratory setting. Practitioners are, however, particularly sensitive to the importance of situation awareness even though researchers find that a clear definition remains elusive (Sarter and Woods, 1991; Woods and Sarter, 2010).

Understanding these attentional dynamics relative to task complexities and how they are affected by computer-based systems is a very important research issue for progress in aiding situation awareness and for safety in supervisory control systems (cf. McRuer et al., (eds) 1992), National Academy of Sciences report on Aeronautical Technologies for the Twenty-First Century, Chapter 11). To meet this research objective we will need to understand more about coordination across human and machine agents, about how to increase the observability of the state and activities of automated systems, and about what are the critical characteristics of displays that integrate multiple sources of data in mentally economical ways.

FAILURES TO REVISE SITUATION ASSESSMENTS: FIXATION OR COGNITIVE LOCKUP

The results of several studies (e.g., De Keyser and Woods, 1990; Cook, McDonald, and Smalhout, 1989; Johnson et al., 1981, 1988; Gaba and DeAnda, 1989; Dunbar, 1995; Klein, Pliske, et al., 2005; Rudolph et al., 2009) strongly suggest that one source of error in dynamic domains is a failure to revise situation assessment as new evidence comes in. Evidence discrepant with the agent's or team's current assessment is missed or discounted or rationalized as not really being discrepant with the current assessment. The operational teams involved in several major accidents seem to have exhibited this pattern of behavior;

examples include the Three Mile Island accident (Kemeny et al., 1979) and the Chernobyl accident.

Many critical real-world human problem-solving situations take place in dynamic, event-driven environments where the evidence arrives over time and situations can change rapidly. Incidents rarely spring full blown and complete; incidents evolve. In these situations, people must amass and integrate uncertain, incomplete, and changing evidence; there is no single well-formulated diagnosis of the situation. Rather, practitioners make provisional assessments and form expectancies based on partial and uncertain data. These assessments are incrementally updated and revised as more evidence comes in. Furthermore, situation assessment and plan formulation are not distinct sequential stages, but rather they are closely interwoven processes with partial and provisional plan development and feedback leading to revised situation assessments (Woods and Roth, 1988; Klein et al., 1993; Woods and Hollnagel, 2006).

In psychological fixations (also referred to as cognitive lockup and cognitive hysteresis), the initial situation assessment tends to be appropriate, in the sense of being consistent with the partial information available at that early stage of the incident. As the incident evolves, however, people fail to revise their assessments in response to new evidence, evidence that indicates an evolution away from the expected path. The practitioners become fixated on an old assessment and fail to revise their situation assessment and plans in a manner appropriate to the data now present in their world. Thus, a fixation occurs when practitioners fail to revise their situation assessment or course of action and maintain an inappropriate judgment or action in the face of opportunities to revise.

Several criteria are necessary to describe an event as a fixation. One critical feature is that there is some form of persistence over time in the behavior of the fixated person or team. Second, opportunities to revise are cues, available or potentially available to the practitioners, that could have started the revision process if observed and interpreted properly. In part, this feature distinguishes fixations from simple cases of inexperience, lack of knowledge, or other problems that impair error detection and recovery (Cook et al., 1989). As with the label "loss of situation awareness," the problem is to define a standard to use to determine what cue or when a cue should alert the practitioners to the discrepancy between the perceived state of the world and the actual state of the world. There is a great danger of falling into the hindsight bias when evaluating after the fact whether a cue "should" have alerted the problem solvers to the discrepancy. The basic defining characteristic of fixations is that the immediate problem-solving context has biased the practitioners in some direction. In naturally occurring problems, the context in which the incident occurs and the way the incident evolves activates certain kinds of knowledge as relevant to the evolving incident. This knowledge, in turn, affects how new incoming information is interpreted. After the fact or after the correct diagnosis has been pointed out, the solution seems obvious, even to the fixated person or team.

De Keyser and Woods (1990) describe several patterns of behavior that have been observed in cases of practitioner fixation. In the first one, "everything but that," the operators seem to have many hypotheses in mind, but never entertain the correct one. Their external behavior looks incoherent because they are often jumping from one action to another one without any success. The second one is the opposite: "this and nothing else." The practitioners are stuck on one strategy, one goal, and they seem unable to shift or to consider other possibilities. One can observe a great deal of persistence in their behavior in this kind of case; for example, practitioners may repeat the same action or recheck the same data channels several times. This pattern is easy to see because of the unusual level of repetitions despite an absence of results. The practitioners often detect the absence of results themselves but without any change in strategy. A third pattern is "everything is O.K." In this case, the practitioners do not react to the change in their environment. Even if there are multiple cues and evidence that something is going wrong, they do not seem to take these indicators at face value. They seem to discount or rationalize away indications that are discrepant with their model of the situation. On the other hand, one must keep in mind the demands of situation assessment in complex fields of practice. For example, some discrepant data actually may be red herrings or false alarms which should be discounted for effective diagnostic search (e.g., false or nuisance alarms can be frequent in many systems). This is essentially a strategic dilemma in diagnostic reasoning, the difficulty of which depends in part on the demands of problems and on the observability of the processes in question.

There are certain types of problems that may encourage fixations by mimicking other situations, in effect, leading practitioners down a garden path (Johnson et al., 1988; Johnson, Jamal, and Berryman, 1991; Johnson, Grazioli, Jamal, and Zualkernan, 1992). In garden path problems "early cues strongly suggest [plausible but] incorrect answers, and later, usually weaker cues suggest answers that are correct" (Johnson, Moen, and Thompson, 1988). It is important to point out that the erroneous assessments resulting from being led down the garden path are not due to knowledge factors. Rather, they seem to occur because "a problem-solving process that works most of the time is applied to a class of problems for which it is not well suited" (Johnson et al., 1988). This notion of garden path situations is important because it identifies a task genotype in which people become susceptible to fixations (McGuirl et al., 2009). The problems that occur are best attributed to the interaction of particular environmental (task) features and the heuristics people apply (local rationality given difficult problems and limited resources), rather than to any particular bias or problem in the strategies used. The way that a problem presents itself to practitioners may make it very easy to entertain plausible but in fact erroneous possibilities.

Diagnostic problems fraught with inherent uncertainties are common in complex fields of practice (Woods and Hollnagel, 2006). As a result, it may be necessary for practitioners to entertain and evaluate what turn out later to be erroneous assessments. Problems arise when the revision process breaks down and the practitioner becomes fixated on an erroneous assessment, missing, discounting or re-interpreting discrepant evidence (see Johnson et al., 1988; Roth, Woods, and Pople, 1992; McGuirl et al., 2009 for analyses of performance in garden path incidents). What is important is the process of error detection and recovery which fundamentally involves searching out and evaluating discrepant evidence to keep up with a changing incident.

Several cognitive processes involved in attentional dynamics which may give rise to fixation:

○ breakdowns in shifting or scheduling attention as the incident unfolds;

- factors of knowledge organization and access that make critical knowledge inert;
- difficulties calling to mind alternative hypotheses that could account for observed anomalies – problems in the processes underlying hypothesis generation;
- problems in strategies for situation assessment (diagnosis) given the probability of multiple factors, for example how to value parsimony (single factor assessments) versus multi-factor interpretations.

Fixation may represent the downside of normally efficient and reliable cognitive processes involved in diagnosis and disturbance management in dynamic contexts. Although fixation is fundamentally about problems in attentional dynamics, it may also involve inert knowledge (failing to call to mind potentially relevant knowledge such as alternative hypotheses) or strategic factors (tradeoffs about what kinds of explanations to prefer).

It is clear that in demanding situations where the state of the monitored process is changing rapidly, there is a potential conflict between the need to revise the situation assessment and the need to maintain coherence. Not every change is important; not every signal is meaningful. The practitioner whose attention is constantly shifting from one item to another may not be able to formulate a complete and coherent picture of the state of the system. For example, the practitioner in Case 6.1 was criticized for failing to build a complete picture of the patient's changing physiological state. Conversely, the practitioner whose attention does not shift may miss cues and data that are critical to updating the situation assessment. This latter condition may lead to fixation. How practitioners manage this conflict is largely unstudied.

Given the kinds of cognitive processes that seem to be involved in fixation, there are a variety of techniques that, in principle, may reduce this form of breakdown. Research consistently shows that revising assessments successfully requires a new way of looking at previous facts (Woods et al., 1987; Patterson et al., 2001). We provide this "fresh" point of view: (a) by bringing in people new to the situation, (b) through interactions across diverse groups with diverse knowledge and tools, (c) through new visualizations which capture the big picture and re-organize data into different perspectives. The latter is predicated on the fact that poor feedback about the state and behavior of the monitored process, especially related to goal achievement, is often implicated in fixations and failures to revise. Thus, one can provide practitioners with new kinds of representations about what is going on in the monitored process (cf. Woods et al., 1987 for examples from nuclear power which tried this in response to the Three Mile Island accident).

Note how avoiding fixation and improving the ability to revise assessments reveals the multi-agent nature of cognitive activities in the wild. One changes the architecture of the distributed system to try to ensure a fresh point of view, that is, one that is unbiased by the immediate context. In these distributed system architectures some members or teams develop their views of the evolving situation separately from others. As a result, one person or group can cross-checks the assessments developed by others. These collaborative inter-changes then can generate fresh points of view or produce challenges to basic assumptions. For example, this cross checking process is an important part of how NASA mission control responds to anomalies (Watts-Perotti and Woods, 2009; see also, Patterson, Roth et al., 2004; Patterson, Woods, et al., 2007; Klein, Feltovich, et al., 2005; Patterson et al., 2007).

8
GOAL CONFLICTS

STRATEGIC FACTORS

A third set of factors at work in distributed cognitive systems is strategic in nature. People have to make tradeoffs between different but interacting or conflicting goals, between values or costs placed on different possible outcomes or courses of action, or between the risks of different errors (Brown, 2005a, 2005b; Woods, 2006; Hollnagel, 2009). They must make these tradeoffs while facing uncertainty, risk, and the pressure of limited resources (e.g., time pressure; opportunity costs).

CASE 8.1   BUSY WEEKEND OPERATING SCHEDULE

On a weekend in a large tertiary care hospital, the anesthesiology team (consisting of four physicians, three of whom are residents in training) was called on to perform anesthetics for an in vitro fertilization, a perforated viscus, reconstruction of an artery of the leg, and an appendectomy, in one building, and one exploratory laparotomy in another building. Each of these cases was an emergency, that is, a case that cannot be delayed for the regular daily operating room schedule. The exact sequence in which the cases were done depended on multiple factors. The situation was complicated by a demanding nurse who insisted that the exploratory laparotomy be done ahead of other cases. The nurse was responsible only for that single case; the operating room nurses and technicians for that case could not leave the hospital until the case had been completed.

The surgeons complained that they were being delayed and their cases were increasing in urgency because of the passage of time. There were also some delays in preoperative preparation of some of the patients for surgery. In the primary operating room suites, the staff of nurses and technicians were only able to run two operating rooms simultaneously.

> The anesthesiologist in charge was under pressure to attempt to overlap portions of procedures by starting one case as another was finishing so as to use the available resources maximally. The hospital also served as a major trauma center which means that the team needed to be able to start a large emergency case with minimal (less than 10 minutes) notice. In committing all of the residents to doing the waiting cases, the anesthesiologist in charge produced a situation in which there were no anesthetists available to start a major trauma case. There were no trauma cases and all the surgeries were accomplished. The situation was so common in the institution that it was regarded by many as typical rather than exceptional.
>
> In this incident, the anesthesiologist in charge committed all of his available resources, including himself, to doing anesthesia. This effectively eliminated the in-charge person's ability to act as a buffer or extra resource for handling an additional trauma case or a request from the floor. In the institution where the incident occurred, the anesthetist in charge on evenings and weekends determines which cases will start and which ones will wait. Being in charge also entails handling a variety of emergent situations in the hospital. These include calls to intubate patients on the floors, requests for pain control, and handling new trauma cases. The anesthesiologist in charge also serves as a backup resource for the operations in progress, sometimes described as an "extra pair of hands". For a brief period of time, the system was saturated with work. There were no excess resources to apply to a new emergency. The anesthesiologist in charge resolved the conflict between the demand for use of all the resources at his command and the demand for preservation of resources for use in some emergency in favor of using all the resources to get urgent work completed. This was a gamble, a bet that the work at hand would continue to be manageable until enough was completed to restore some resource excess that would provide a margin for handling unforeseen (and unforeseeable) emergencies. These factors were not severe or particularly unusual. Rather, they represented the normal functioning of a large urban hospital. The decision to proceed along one pathway rather than another reflects on the way that the practitioner handled the strategic factors associated with his work that day.

One remarkable aspect of this incident is that it was regarded as unremarkable by the participants. These kinds of scheduling issues recur and are considered by many to be simply part of the job. There were strong incentives to commit the resources, but also incentives to avoid that commitment. Factors that played a role in the anesthetist's decision to commit all available resources included the relatively high urgency of the cases, the absence of a trauma alert (indication that a trauma patient was in route to the hospital), the time of day (fairly early; most trauma is seen in the late evening or early morning hours), and the pressure from surgeons and nurses.

Another reason for committing the resources seems almost paradoxical. The practitioner needed to commit resources in order to free them for work. By completing cases as early as possible in the day, he was providing resources for use in the late evening, when trauma operations seemed more likely. The resources at hand were available but evanescent; there was no way to 'bank' them against some future need. Refusing to use them now did not assure that they would be available in the future. Rather this would actually reduce the likelihood that they would become available. Indeed, it seemed clear to the practitioner that it was desirable to get things done in order to be prepared for the future. The closer circumstances came to saturating the group's ability to handle cases, the greater was urgency to complete cases already in progress or pending in the near future.

One can argue (and there were discussions amongst practitioners at the time) that committing all the resources was a risky course. It is not the ultimate resolution of this debate that matters here. Rather what this incident shows is the way that the domain of practice confronts practitioners with the need to act decisively in the midst of competing demands and in an environment rife with uncertainty. The incident demonstrates not that the practitioner was either a risk taker or risk averse. There is generally no risk free way to proceed in these sorts of domains. Rather, all the possible ways of proceeding involve exposure to different sets of risks. The decision to proceed in one way or some other way is the result of coping with multiple, conflicting goals and demands, trading off aspects and elements of one against elements and aspects of the other.

This incident typifies the goal interactions and dilemmas that arise in cognitive work. These are the ways in which people accommodate a variety of problematic issues: the different values of possible outcomes; the implications of the costs of the different courses of action; the consequences of various kinds of possible failure. Coping with these issues is fundamentally a strategic process, one that involves assessments and comparisons of different possible actions. It is especially relevant that these assessments and comparisons take place under time pressure and in the face of uncertainty. The central theme of this chapter is the role of dilemmas and goal conflicts in complex system failures. Focusing on the multiple goals that play out in sharp end practice inevitably force us to look outward to the larger organizational contexts which shapes the nature of practice.

MULTIPLE, CONFLICTING GOALS

Multiple, simultaneously active goals are the rule rather than the exception for virtually all domains in which expertise is involved. Practitioners must cope with the presence of multiple goals, shifting between them, weighing them, choosing to pursue some rather than others, abandoning one, embracing another. Many of the goals encountered in practice are implicit and unstated. Goals often conflict. Sometimes these conflicts are easily resolved in favor of one or another goal, sometimes they are not. Sometimes the conflicts are direct and irreducible, for example when achieving one goal necessarily precludes achieving another one. But there are also intermediate situations, where several goals may be partially satisfied simultaneously. An adequate analysis of real world incidents requires explicit description of the interacting goals and the ways in which practitioners assessed and compared them.

Perhaps the most common hazard in the analysis of incidents is the naïve assessment of the strategic issues that confront practitioners. It is easy, especially after accidents, to simplify the situation facing practitioners in ways that ignore the real pressures and demands placed on them, that is to attempt to resolve the goal conflicts that actually exist by the convenient fiat of discarding some of them as insignificant or immaterial.

Some goal conflicts are inherently technical; they arise from intrinsic technical nature of a domain. In the case of the anesthetized patient, a high blood pressure works to push blood through the coronary arteries and improve oxygen supply to the heart muscle but, because increased blood pressure adds to cardiac work, a low blood pressure should reduce cardiac work. The appropriate blood pressure target adopted by the anesthetist depends in part on the practitioner's strategy, the nature of the patient, the kind of surgical procedure, the circumstances within the case that may change (e.g., the risk of major bleeding), and the negotiations between different people in the operating room team (e.g., the surgeon who would like the blood pressure kept low to limit the blood loss at the surgical site). This is a simple example because the practitioner is faced with a continuum of possible blood pressures. It is easy to imagine a momentary "optimal" blood pressure in between the extremes in the attempt to manage both goals.

But there are parallel examples of conflicting goals that are not so clearly technical in character (Brown, 2005a). For example, a very high level goal (and the one most often explicitly acknowledged) in anesthesiology is to preserve the patient's life. If this were the only active goal, then we might expect the practitioner to behave in a certain way, studiously avoiding all risks, refraining from anything that might constitute, even in hindsight, exposure to hazard. But this is not the only goal; there are others. These include reducing costs, avoiding actions that would lead to lawsuits, maintaining good relations with the surgical service, maintaining resource elasticity to allow for handling unexpected emergencies, providing pain relief, and many others. Incidents and accidents have a way of making some goals appear to have been crucial and others irrelevant or trivial, but this hindsight actually blinds us to the relevance of goals before the event. Our knowledge of outcome makes it impossible for us to weigh the various goals as if we were the practitioners embedded in the situation. But in order to understand the nature of behavior, it is essential to bring these various goals into focus. The incident that begins this chapter shows how sterile and uninformative it is to cast the situation confronting a practitioner as dominated by a single, simple goal. Preserving the patient's life is indeed a goal, and a high level one, but this practitioner is dealing with multiple patients – even including hypothetical ones, such as the potential trauma patient. How is he to assess the relative acuity and hazard of the different situations with which he must cope?

Individual goals are not the source of the conflicts; conflicts arise from the relationships between different goals. In the daily routine, for example, the goal of discovering important medical conditions with anesthetic implications before the day of surgery may drive the practitioner to seek more information about the patient. Each hint of a potentially problematic condition provides an incentive for further tests that incur costs (e.g., the dollar cost of the tests, the lost opportunity cost when a surgical procedure is canceled and the operating room goes unused for that time, the social costs disgruntled surgeons). The goal of minimizing costs, in contrast, provides an incentive for a minimal preoperative testing and the use of same-day surgery. The external pressures for highly efficient performance are in favor of limiting the preoperative evaluation (Woods, 2006). But failing to acquire information may reduce the ability of the anesthesiologist to be prepared for events during surgery and contribute to the evolution of a disaster.

To take another example of conflicting goals, consider the in-flight modification of flight plans in commercial aviation. At first glance it seems a simple thing to conclude that flight plans should be modified in the interests of "safety" whenever the weather is bad on the path ahead. There are, however, some goals that need to be included in the decision to modify the plan. Avoiding passenger discomfort and risk of injury from turbulence are goals that are, in this case, synonymous with "safety". But there are other goals active at that same time including the need to minimize fuel expenditure, and the need to minimize the difference between the scheduled arrival time and actual arrival time. These latter goals are in conflict with the former ones (at least in some situations) because achieving one set risks failing to achieve the other. The effect of these competing goals on pilots and ground controllers is complex and forces tradeoffs between the goals by pilots and ground controllers.

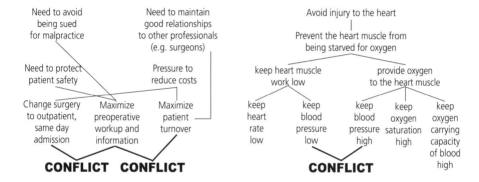

Figure 8.1 Conflicting goals in anesthesia. Some goals, for example acquiring knowledge of underlying medical conditions and avoiding successful lawsuits, create an incentive for preoperative testing. The goal of reduced costs creates an incentive for the use of outpatient surgery and limited preoperative evaluation

Because goal conflicts become easier to see at the sharp end of the system, the tendency is to regard practitioners themselves as the sources of conflict. It should be clear, however, that organizational factors at the blunt end of systems shape the world in which practitioners work. The conflicting or inherently contradictory nature of goals that practitioners confront are usually derived from the organizational factors such as management policies, the need to anticipate legal liability, regulatory guidelines, and economic factors. Competition between goals generated at the organizational level was an important factor in the breakdown of safety barriers in the system for transporting oil through Prince William Sound that preceded the *Exxon Valdez* disaster (National Transportation Safety Board, 1990). Even when the practitioners seem to be more actively the "cause" of accidents, the conflicted nature of the goals they confront is a critical factor.

In cases where the investigations are more thorough and deep, this pattern can be seen more clearly. Aviation accidents, with their high public profile, are examples. There have been several crashes where, in hindsight, crews encountered a complex web of conflicted goals that led to attempted takeoffs with ice on the wings (National Transportation Safety Board, 1993). In one such case, the initial "investigation" of the accident focused narrowly on practitioner error and the second story became apparent only after a national judicial inquiry (Moshansky, 1992).

The goals that drive practitioner behavior are not necessarily those of written policies and procedures. Indeed, the messages received by practitioners about the nature of the institution's goals may be quite different from those that management acknowledges. Many goals are implicit and unstated. This is especially true of the organizational guidance given to practitioners. For example, the Navy sent an implicit but very clear message to its commanders by the differential treatment it accorded to the commander of the Stark following that incident (U.S. House of Representatives Committee on Armed Services, 1987) as opposed to the Vincennes following that incident (U.S. Department of Defense, 1988; Rochlin, 1991). These covert factors are especially insidious because they shape and constrain behavior and, in politicized and risky settings, because they are difficult to acknowledge. After accidents we can readily obtain the formal policies and written procedures but it is more difficult to capture the web of incentives and imperatives communicated to workers by management in the period that precedes accidents (Stephenson et al., 2000; CAIB, 2003; Woods, 2005).

Coping with Multiple Goals

The dilemmas that inevitably arise out of goal conflicts may be resolved in a variety of ways. In some cases, practitioners may search deliberately for balance between the competing demands. In doing so their strategies may be strong and robust, brittle (work well under some conditions but are vulnerable given other circumstances), or weak (highly vulnerable to breakdown). They may also not deliberate at all but simply apply standard routines. In any case, they either make or accept tradeoffs between competing goals. In the main, practitioners are successful in the effort to strike a balance between the multiple coals present in the domain of practice.

In general, outsiders pay attention to practitioners' coping strategies only after failure, when such processes seem awkward, flawed, and fallible. It is easy for post-incident evaluations to say that a human error occurred. It is obvious after accidents that practitioners should have delayed or chosen some other means of proceeding that would have avoided (what we now know to be) the complicating factors that precipitated an accident. The role of goal conflicts arising from multiple, simultaneously active goals may never be noted. More likely, the goal conflicts may be discounted as requiring no effort to resolve, where the weighting of goals appears in retrospect to "have been obvious" in ways that require no discussion.

NASA's "Faster, Better, Cheaper" organizational philosophy in the late 1990s epitomized how multiple, contradictory goals are simultaneously present and active in complex systems (Woods, 2006). The loss of the Mars Climate Orbiter and the Mars Polar Lander in 1999 were ascribed in large part to the irreconcilability of the three goals (faster and better and cheaper), which drove down the cost of launches, made for shorter, aggressive mission schedules, eroded personnel skills and peer interaction, limited time, reduced the workforce, and lowered the level of checks and balances normally found (Stephenson, 2000). People argued that NASA should pick any two from the three goals. Faster and cheaper would not mean better. Better and cheaper would mean slower. Faster and better would be more expensive. Such reduction, however, obscures the actual reality facing operational personnel in safety-critical settings. These people are there to pursue all three goals simultaneously – fine-tuning their operation, as Starbuck and Milliken (1988) said, to "render it less redundant, more efficient, more profitable, cheaper, or more versatile" (p. 323).

CASE 8.2 SPACE SHUTTLE *COLUMBIA* EXTERNAL TANK MAINTENANCE

The 2003 space shuttle *Columbia* accident focused attention on the maintenance work that was done on the Shuttle's external fuel tank, once again revealing the differential pressures of having to be safe and getting the job done (better, but also faster and cheaper). A mechanic working for the contractor, whose task it was to apply the insulating foam to the external fuel tank, testified that it took just a couple of weeks to learn how to get the job done, thereby pleasing upper management and meeting production schedules. An older worker soon showed him how he could mix the base chemicals of the foam in a cup and brush it over scratches and gouges in the insulation, without reporting the repair. The mechanic soon found himself doing this hundreds of times, each time without filling out the required paperwork. Scratches and gouges that were brushed over with the mixture from the cup basically did not exist as far as the organization was concerned. And those that did not exist could not hold up the production schedule for the external fuel tanks. Inspectors often did not check. A company program that once had paid workers hundreds of dollars for finding defects had been watered down, virtually inverted by incentives for getting the job done now.

Goal conflicts between safer, better, and cheaper were reconciled by doing the work more cheaply, superficially better (brushing over gouges), and apparently without cost to safety. As in most operational work, the distance between formal, externally dictated logics of action and actual work is bridged with the help of those who have been there before, who have learned how to get the job done (without apparent safety consequences), and who are proud to share their professional experience with younger, newer workers. Actual practice settles at a distance from the formal description of the job. Informal networks may characterize work, including informal hierarchies of teachers and apprentices and informal documentation of how to actually get work done. The notion of an incident, of something that was worthy of reporting (a defect) becomes blurred against a background of routine nonconformity. What was normal versus what was problem was no longer so clear.

RESPONSIBILITY-AUTHORITY DOUBLE BINDS

A crucial dilemma that plays a role in incidents and accidents involves the relationship between authority and responsibility. *Responsibility-authority double binds* are situations in which practitioners have the responsibility for the outcome but lack the authority to take the actions they see as necessary. In these situations practitioners' authority to act is constrained while they remain vulnerable to penalties for bad outcomes. This may arise from vesting authority in external agents via control at a distance (e.g., via the regimentation "just follow the procedures") or the introduction of machine-cognitive agents that automatically diagnose situations and plan responses.

People working cooperatively (and effectively) tend to pass authority with responsibility together in advisory interactions, that is, in most cases where small groups work together, responsibility and authority stay together (Hoffman et al., 2009). The research results regarding the consequences of dividing responsibility from authority are limited but consistent. Splitting authority and responsibility appears to have poor consequences for the ability of operational systems to handle variability and surprises that go beyond pre-planned routines (Roth et al., 1987; Hirschhorn, 1993). Billings (1991) uses this idea as the fundamental premise of his approach to develop a human-centered automation philosophy – if people are to remain responsible for safe operation, then they must retain effective authority. Automation that supplants rather than assists practitioners violates this fundamental premise.

The paradox here is that attribution of accident cause to human (operator) error often leads to the introduction of organizational change that worsens authority-responsibility double binds. Seeing the operators as the source of failure provides the incentive to defend the system against these acts, usually through regimentation. This effectively leads to a loss of authority. But the complexity of the worlds of practice means that attempts at complete regimentation produce more brittle work systems (but makes it easier to diagnose human error in terms of failures to follow rules).

One consequence of the Three Mile Island nuclear reactor accident was a push by the Nuclear Regulatory Commission for utility companies to develop more detailed and comprehensive work procedures and to ensure that operators followed these procedures exactly. This policy appeared to be a reasonable approach to increase safety. However, for the people at the sharp end of the system who actually did things, strictly following the procedures posed great difficulties. The procedures were inevitably incomplete, and sometimes contradictory (dilemmas about what it meant to follow procedures in complex dynamic abnormal situations arose in a variety of incidents; see Roth et al., 1992, for a study of a simulated emergency where the procedure was incomplete). Then too, novel circumstances arose that were not anticipated in the work procedures. The policy inevitably leads to situations where there is a "double bind" because the people would be wrong if they violated a procedure even though it could turn out to be an inadequate procedure, and they would be wrong if they followed a procedure that turned out to be inadequate.

In some situations, if they followed the standard procedures strictly the job would not be accomplished adequately; if they always waited for formal permission to deviate from standard procedures, throughput, and productivity would be degraded substantially. If they deviated and it later turned out that there was a problem with what they did (e.g., they did not adapt adequately), it could create re-work or safety or economic problems. The double bind arises because the workers are held responsible for the outcome (the poor job, the lost productivity, or the erroneous adaptation); yet they did not have authority for the work practices because they were expected to comply exactly with the written procedures. As Hirschhorn says:

> They had much responsibility, indeed as licensed professionals many could be personally fined for errors, but were uncertain of their authority. What freedom of action did they have, what were they responsible for? This gap between responsibility and authority meant that operators and their supervisors felt accountable for events and actions they could neither influence nor control. (Hirschhorn, 1993)

Workers coped with the double bind by developing a "covert work system" that involved, as one worker put it, "doing what the boss wanted, not what he said" (Hirschhorn, 1993). There were channels for requesting changes to problems in the procedures, but the process was cumbersome and time-consuming. This is not surprising since, if modifications are easy and liberally granted, then it may be seen as undermining the policy of strict procedure-following. Notice how the description of this case may fit many different domains (e.g., the evolving nature of medical practice).

The design of computer-based systems has also been shown to be a factor that can create authority-responsibility double binds (Roth et al., 1987). Consider a traditional artificial intelligence based expert system that solves problems on its own, communicating with the operator via a question and answer dialogue. In this approach to assistance, the machine is in control of the problem; the system is built on the premise that the expert system can solve the problem on its own if given the correct data. The human's role is to serve as the system's interface to the environment by providing it with the data to solve the problem. If the human practitioners are to do any problem solving, it is carried out in parallel, independent of the interaction with the intelligent system. Results indicate that this prosthesis form of interaction between human and intelligent system is very brittle in the face of complicating factors (Roth et al., 1987). Again, the need to cope with novel situations, adapt to special conditions or contexts, recover from errors in following the instructions, or cope with bugs in the intelligent system itself requires a robust cognitive system that can detect and recover from error.

The crux of the problem in this form of cooperation is that the practitioner has responsibility for the outcome of the diagnosis, but the machine expert has taken over effective authority through control of the problem-solving process. Note the double bind practitioners are left in, even if the machine's solution is disguised as only "advice" (Roth et al., 1987; Woods, 1991). In hindsight, practitioners would be wrong if they failed to follow the machine's solution and it turned out to be correct, even though machine can err in some cases. They would be wrong if they followed the machine's "advice" in those cases where it turned out the machine's solution was inadequate. They also would be wrong if they were correctly suspicious of the machine's proposed solution, but failed to handle the situation successfully through their own diagnosis or planning efforts (see Part V on how knowledge of outcome biases evaluation of process). The practitioners in the evolving problem do not have the advantage of knowledge of the eventual outcome when they must evaluate the data at hand including the uncertainties and risks.

Instructions, however elaborate, regardless of medium (paper- or computer-based), and regardless of whether the guidance is completely pre-packaged or partially generated "on-the-fly" by an expert system, are inherently brittle when followed rotely. Brittleness means that it is difficult to build in mechanisms that cope with novel situations, adapt to special conditions or contexts, or recover from errors in following the instructions or bugs in the instructions themselves (e.g., Brown, Moran, and Williams, 1982; Woods et al., 1987; Herry, 1987; Patterson et al., 2010). As Suchman (1987) has put it, "plans are [only] resources for action."

When people use guidance to solve problems, erroneous actions fall into one of two general categories (Woods et al., 1987; Woods and Shattuck, 2000):

1. rote-rule following persists in the face of changing circumstances that demand adaptation,
2. the people correctly recognize that standard responses are inadequate to meet operational goals given the actual circumstances, but fail to adapt the pre-planned guidance effectively (e.g., missing a side effect).

For example, studies of nuclear power plant operators responding to simulated and to actual accident conditions with paper-based instructions found that operator performance problems fell into one or the other of the above categories (Woods et al., 1987). If practitioners (those who must do something) are held accountable for both kinds of "error" – those where they continue to rotely follow the rules in situations that demand adaptation and those where they erroneously adapt – then the practitioners are trapped in a double bind.

Following instructions requires actively filling in gaps based on an understanding of the goals to be achieved and the structural and functional relationships between objects referred to in the instructions. For example, Smith and Goodman (1984) found that more execution errors arose in assembling an electrical circuit when the instructions consisted exclusively of a linear sequence of steps to be executed, than when explanatory material related the instruction steps to the structure and function of the device. Successful problem solving requires more than rote instruction following; it requires understanding how the various instructions work together to produce intended effects in the evolving problem context.

While some of the problems in instruction following can be eliminated by more carefully worded, detailed, and explicit descriptions of requests, this approach has limitations. Even if, in principle, it were possible to identify all sources of ambiguity and craft detailed wording to avoid them, in practice the resources required for such extensive fine tuning are rarely available. Furthermore, the kinds of literal elaborate statements that would need to be developed to deal with exceptional situations are likely to obstruct the comprehension and execution of instructions in the more typical and straightforward cases (for example, in one aviation incident the crew used about 26 different procedures; see Part IV for more on this incident).

Attempts to eliminate all sources of ambiguity are fundamentally misguided. Examination of language use in human-human communication reveals that language is inherently underspecified, requires the listener (or reader) to fill in gaps based on world knowledge, and to assess and act on the speaker's (writer's) intended goals rather than his literal requests (Suchman, 1987). Second, a fundamental competency in human-human communication is the detection and repair of communication breakdowns (Suchman, 1987; Klein, Feltovish, et al., 2005). Again, error recovery is a key process. In part, this occurs because people build up a shared frame of reference about the state of the world and about what are meaningful activities in the current context.

Whenever organizational change or technology change occurs, it is important to recognize that these changes can sharpen or lessen the strategic dilemmas that arise in operations and change how practitioners negotiate tradeoffs in context. In designing high reliability systems for fields of activity with high variability and potential for surprise (Woods and Hollnagel, 2006), one cannot rely just on rotely followed pre-planned routines (even with a tremendous investment in the system for producing and changing the routines). Nor can one rely just on the adaptive intelligence of people (even with a tremendous investment in the people in the system). Distributed cognitive system design should instead focus on how to coordinate pre-planned routines with the demands for adaptation inherent in complex fields of activity (Woods, 1990a). The history of mission control is a good illustration of the coordination of these two types of activity in pace with the varying rhythms of the field of practice (e.g., Murray and Cox, 1989; Watts-Perotti and Woods, 2009).

It is tempting to oversimplify dilemmas as a tradeoff between safety and economy. But dilemmas and goal conflict that confront practitioners are more intimately connected to the details of sharp end practice. During the management of faults and failures, for example, there is a tradeoff with respect to when to commit to a course of action (Woods and Hollnagel, 2006). Practitioners have to decide when to decide. Should they to take corrective action early in the course of an incident with limited information? Should they instead delay their response and wait for more data to come in or ponder additional alternative hypotheses? Act too early, when the diagnosis of the failure is uncertain, and there is a risk of making the situation worse through the wrong action. Act too late and the failure may have progressed to the point where the consequences have increased in scope or changed in kind or even become irremediable.

A dramatic example of these concerns occurred during the Apollo 13 mission. An explosion in the cryogenics systems led to the loss of many critical spacecraft functions and threatened the loss of the spacecraft and crew (see Murray and Cox, 1989).

Lunney [the Flight Director] was persistent because the next step they were contemplating was shutting off the reactant valve in Fuel Cell 1, as they had done already in Fuel Cell 3. If they shut it off and then came up with a ... solution that suddenly got the O_2 pressures back up, the door would still be closed on two-thirds of the C.S.M's power supply. It was like shooting a lame horse if you were stranded in the middle of a desert. It might be the smart thing to do, but it was awfully final. Lunney, like Kranz before him, had no way of knowing that the explosion had instantaneously closed the reactant valves on both fuel cells 1 and 3. At ten minutes into his shift, seventy-nine minutes after the explosion, Lunney was close to exhausting the alternatives.

"You're ready for that now, sure, absolutely, EECOM [the abbreviation for one of the flight controller positions]?"

"That's it, Flight."

"It [the oxygen pressure] is still going down and it's not possible that the thing is sorta bottoming out, is it?"

"Well, the rate is slower, but we have less pressure too, so we would expect it to be a bit slower."

"You are sure then, you want to close it?"

"Seems to me we have no choice, Flight."

"Well . . ."

Burton, under this onslaught, polled his back room one last time. They all agreed. "We're go on that, Flight."

"Okay, that's your best judgment, we think we ought to close that off, huh?"

"That's affirmative."

Lunney finally acquiesced. "Okay. Fuel Cell 1 reactants coming off."

It was uncharacteristic behavior by Lunney – "stalling," he would later call it. "Just to be sure. Because it was clear that we were at the ragged edge of being able to get this thing back. ... That whole night, I had a sense of containing events as best we could so as not to make a serious mistake and let it get worse."

The role of both formal rules and rules "of thumb" is centrally concerned with the tradeoffs by practitioners. Practitioners frequently trade off between acting based on operational rules or based on reasoning about the case itself (cf. Woods et al., 1987). The issue, often non-trivial, is whether the standard rules apply to a particular situation. When some additional factor is present that complicates the textbook scenario the practitioner must decide whether to use the standard plan, adapt the standard plan in some way, or abandon the standard plan and formulate a new one (Woods and Shattuck, 2000; Woods and Hollnagel, 2006, Chapter 8; Watts-Perotti and Woods, 2009).

The precise criteria for evaluating these different tradeoffs may not be set by a conscious process or an overt decision made by individuals. It is more likely that they are established as emergent properties of either small groups or larger organizations. The criteria may be fairly labile and susceptible to influence, or they may be relatively stable and difficult to change.

The discussion of rules can be extended to a discussion of the coordination among agents in the distributed cognitive system (Woods and Hollnagel, 2006, Chapter 12). Such agents can be people working together or, increasingly, can be people working with information technology (Roth, Bennett, and Woods, 1987). What if the practitioner's evaluation is different from that made by a computer agent? When should the machine's guidance be sufficient? What is enough evidence that the machine is wrong to justify disregarding the machine's evaluation and proceeding along other lines?

In hindsight, practitioner's choices or actions can often seem to be simple blunders. Indeed, most of the media reports of human error in aviation, transportation, medicine, are tailored to emphasize the extreme nature of the participants' behavior. But a more careful assessment of the distributed system may reveal that goals conflicted or other forms of dilemmas arose. Behavior in the specific incident derives from how the practitioners set their tradeoff criteria across different kinds of risks from different kinds of incidents that could occur. Because incidents usually are evaluated as isolated events, such tradeoffs can appear in hindsight to be unwise or even bizarre. This is mitigated when sets of incidents are used as the basis for examining the larger system (see the discussion of hindsight bias in Part V).

When dilemmas are involved in an incident, changing the behavior of the operational system requires a larger analysis of how one should make the tradeoff. It also involves meaningfully and consistently communicating this policy to the operational system so that practitioners adopt it as their criterion. This may implicitly or explicitly involve the commitment of a different system (an organization's management, an entire industry, a regulatory process). Lanir, Fischhoff, and Johnson, (1988) provide an excellent example through their formal analysis of criteria setting for risk taking within a distributed cognitive system. The danger in missing the role of strategic tradeoffs in producing the observed behavior of operational systems is that the changes made or the messages received by the practitioners can exacerbate dilemmas.

DID THE PRACTITIONERS COMMIT ERRORS?

Given the discussion of cognitive factors (knowledge, mindset, and dilemmas) and of local rationality, let us go back to the three exemplar incidents described earlier in Part III and re-examine them from the perspective of the question: What is human error?

These three incidents are not remarkable or unusual in their own field of activity (urban, tertiary care hospitals) or in other complex domains. In each incident, human performance is closely tied to system performance and to eventual outcome, although the performance of the practitioners is not the sole determinant of outcome. For example, the myocardial infarction following the events of case 6.1. may well have happened irrespective of any actions taken by practitioners. That patient was likely to have an infarction, and it is not possible to say if the anesthetist's actions caused the infarction. The incidents and the analysis of human performance that they prompt (including the role of latent failures in incidents) may make us change our notion of what constitutes a human error.

Arguably, the performance in each exemplar incident is flawed. In retrospect, things can be identified that might have been done differently and which would have forestalled or minimized the incident or its effect. In the myocardial infarction incident, intravascular volume was misassessed and treatment for several simultaneous problems was poorly coordinated. In the hypotension case (case 7.1.), the device setup by practitioners contributed to the initial fault. The practitioners were also unable to diagnose the fault until well after its effects had cascaded into a near crisis. In the scheduling incident (case 8.1.), a practitioner violated policy. He chose one path to meet certain demands, but simultaneously exposed the larger system to a rare but important variety of failure. In some sense, each of the exemplar incidents constitutes an example of human error. Note, however, that each incident also demonstrates the complexity of the situations confronting practitioners and the way in which practitioners adjust their behavior to adapt to the unusual, difficult, and novel aspects of individual situations.

The hypotension incident (case 7.1.) particularly demonstrates the resilience of human performance in an evolving incident. During this incident the physicians engaged successfully in disturbance management to cope with the consequences of a fault (Woods and Hollnagel, 2006). The physicians were unable to identify the exact source of the incident until after the consequences of the fault had ended. However, they were able to characterize the kind of disturbance present and to respond constructively in the face of time pressure. They successfully treated the consequences of the fault to preserve the patient's life. They were able to avoid becoming fixated on pursuing what was the "cause" of the trouble. In contrast, another study of anesthesiologist cognitive activities, this time in simulated difficult cases (Schwid and O'Donnell, 1992), found problems in disturbance management where about one-third of the physicians undertreated a significant disturbance in patient physiology (hypotension) while they over-focused on diagnostic search for the source of the disturbance.

The practitioner was also busy during the myocardial infarction incident, although in this instance the focus was primarily on producing better oxygenation of the blood and control of the blood pressure and not on correcting the intravascular volume. These efforts were significant and, in part, successful. In both incidents 1 and 2, attention is drawn to the practitioner performance by the outcome.

In retrospect some would describe aspects of these incidents as human error. The high urine output with blood high glucose and prior administration of furosemide should have prompted the consideration of low (rather than high) intravascular volume. The infusion devices should have been set up correctly, despite the complicated set of steps involved. The diagnosis of hypotension should have included a closer examination of the infusion devices and their associated bags of fluid, despite the extremely poor device feedback. Each of these conclusions, however, depends on knowledge of the outcome; each conclusion suffers from hindsight bias. To say that something should have been obvious, when it manifestly was not, may reveal more about our ignorance of the demands and activities of this complex world than it does about the performance of its practitioners. It is possible to generate lists of "shoulds" for practitioners in large systems but these lists quickly become unwieldy and, in any case, will tend to focus only on the most salient failures from the most recent accident.

The scheduling incident is somewhat different. In that incident it is clear how knowledge of the outcome biases evaluations of the practitioner performance. Is there an error in case 8.1? If a trauma case had occurred in this interval where all the resources had been committed to other cases, would his decision then be considered an error? On the other hand, if he had delayed the start of some other case to be prepared for a possible trauma case that never happened and the delay contributed to some complication for that patient, would his decision then be considered an error?

Uncovering what is behind each of these incidents reveals the label "human error" as a judgment made in hindsight. As these incidents suggest, human performance is as complex and varied as the domain in which it is exercised. Credible evaluations of human performance must be able to account for all of the complexity that confronts practitioners and the strategies they adopt to cope with that complexity. The term human error should not represent the concluding point but rather the starting point for studies of accident evolution in complex systems.

THE N-TUPLE BIND

The Implications of Local Rationality for Studying Error

One implication of local rationality is that normative procedures based on an ideal or perfect rationality do not make sense in evaluating cognitive systems. Rather, we need to find out what are robust, effective strategies given the resources of the problem solvers (i.e., their strategies, the nature of their working memory and attention, long-term memory organization, and retrieval processes, and so on), and the demands of the problem-solving situation (time pressure, conflicting goals, uncertainty, and so on). Error analyses should be based on investigating demand-resource relationships and mismatches (Rasmussen, 1986). As Simon (1969) points out, "It is wrong, in short, in ignoring the principle of bounded rationality, in seeking to erect a theory of human choice on the unrealistic assumptions of virtual omniscience and unlimited computational power" (p. 202).

Human decision makers generally choose strategies that are relatively efficient in terms of effort and accuracy as task and context demands are varied (Payne et al., 1988; 1990). Procedures that seem "normative" for one situation (non-time constrained) may be severely limited in another problem context (time constrained). In developing standards by which to judge what are effective cognitive processes, one must understand problem solving in context, not in "the abstract." For example, if one were designing a decision aid that incorporated Bayesian inference, one would need to understand the context in which the joint human-machine system functions including such factors as noisy data or time pressure. Fischhoff and Beyth-Marom (1983) point out that applying Bayesian inference in actuality (as opposed to theory) has the following error possibilities: formulation of wrong hypotheses, not correctly eliciting the beliefs and values that need to be incorporated into the decision analysis, estimating or observing prior probabilities and likelihood functions incorrectly, using a wrong aggregation rule or applying the right one incorrectly.

In other words, cognitive strategies represent tradeoffs across a variety of dimensions including accuracy, effort, robustness, risks of different bad outcomes, or the chances for gain from different possible good outcomes. Effective problem-solving strategies are situation specific to some extent; what works well in one case will not necessarily be successful in another. Furthermore, appropriate strategies may change as an incident evolves; for example effective monitoring strategies to detect the initial occurrence of a fault (given normal operations as a background) may be very different from search strategies during a diagnostic phase (Moray, 1984). In understanding these tradeoffs relative to problem demands we can begin to see the idea that expertise and error spring from the same sources.

The assumption of local rationality (Woods and Hollnagel, 2006) – people are doing reasonable things given their knowledge, their objectives, their point of view and limited resources, for example time or workload – points toward a form of error analysis that consists of tracing the problem-solving process to identify points where limited knowledge and limited processing lead to breakdowns. This perspective implies that one must consider what features of domain incidents and situations increase problem demands (Patterson et al., in press).

The incidents described in Part III are exemplars for the different cognitive demands encountered by practitioners who work at the sharp end of large, complex systems, including anesthetists, aircraft pilots, nuclear power plant operators, and others. Each category of cognitive issue (knowledge in context, mindset, strategic factors, and local rationality) plays a role in the conduct of practitioners and hence plays a role in the genesis and evolution of incidents. The division of cognitive issues into these categories provides a tool for analysis of human performance in complex domains. The categories are united, however, in their emphasis on the conflicts present in the domain. The conflicts exist at different levels and have different implications, but the analysis of incidents depends in large part on developing an explicit description of the conflicts and the way in which the practitioners deal with them.

Together the conflicts produce a situation for the practitioner that appears to be a maze of potential pitfalls. This combination of pressures and goals in the work environment is what can be called the n-tuple bind (Cook and Woods, 1994). This term derives from the mathematical concept of a series of numbers required to define an arbitrary point in an *n*-dimensional space. The metaphor here is one of a collection of factors that occur simultaneously within a large range of dimensions, that is, an extension of the notion of a double bind. The practitioner is confronted with the need to choose a single course of action from myriad possible courses. How to proceed is constrained by both the technical characteristics of the domain and the need to satisfy the "correct" set of goals at a given moment, chosen from the many potentially relevant ones. This is an example of an over-constrained problem, one in which it is impossible to maximize the function or work product on all dimensions simultaneously. Unlike simple laboratory worlds with a best choice, real complex systems intrinsically contain conflicts that must be resolved by the practitioners at the sharp end. Retrospective critiques of the choices made in system operation will always be informed by hindsight. For example, if the choice is between obtaining more information about cardiac function or proceeding directly to surgery with a patient who has soft signs of cardiac disease, the outcome will be a potent determinant of the "correctness" of the decision. Proceeding with undetected cardiac disease may lead to a bad outcome (although this is by no means certain), but obtaining the data may yield normal results, cost money, "waste" time, and incur the ire of the surgeon (Woods, 2006). Possessing knowledge of the outcome trivializes the situation confronting the practitioner and makes the "correct" choice seem crystal clear.

This n-tuple bind is most easily seen in case 8.1 where strategic factors dominate. The practitioner has limited resources and multiple demands for them. There are many sources of uncertainty. How long will the in-vitro fertilization take? It should be a short case but may not be. The exploratory laparotomy may be either simple or complex. With anesthetists of different skill levels, who should be sent to the remote location where that case will take place? Arterial reconstruction patients usually have associated heart disease and the case can be demanding. Should he commit the most senior anesthetist to that case? Such cases are also usually long, and committing the most experienced anesthetist will tie up that resource for a long time. What is the likelihood that a trauma case will come during the time when all the cases will be going simultaneously (about an hour)? There are demands from several surgeons for their case to be the next to start. Which case is the most medically important one? The general rule is that an anesthetist has to be available for a trauma; he is himself an anesthetist and could step in but this would leave no qualified individual to go to cardiac arrests in the hospital or to the emergency room. Is it desirable to commit all the resources now and get all of the pending cases completed so as to free the people for other cases that are likely to follow?

It is not possible to measure accurately the likelihood of the various possible events that he considers. As in many such situations in medicine and elsewhere, he is attempting to strike a balance between common but lower consequence problems and rare but higher consequence ones. Ex post facto observers may view his actions as either positive or negative. On the one hand his actions are decisive and result in rapid completion of the urgent cases. On the other, he has produced a situation where emergent cases may be delayed. The outcome influences how the situation is viewed in retrospect.

A critique often advanced in such situations is that "safety" should outweigh all other factors and be used to differentiate between options. Such a critique is usually made by those very far removed the situations that can arise at the sharp end. Safety is not a concrete entity and the argument that one should always choose the safest path misrepresents the dilemmas that confront the practitioner. The safest anesthetic is the one that is not given; the safest airplane is the one that never leaves the ground. All large, complex systems have intrinsic risks and hazards that must be incurred in order to perform their functions, and all such systems have had failures. The investigation of such failures and the attribution of cause by retrospective reviewers are discussed in Part V of this book.

Another aviation example involves de-icing of the wings of planes before winter weather takeoffs at busy airports. The goal of de-icing is to avoid takeoff with ice on the wings. After an aircraft is de-iced, it enters the queue of planes waiting for takeoff. Because the effectiveness of the de-icing agent degrades with time, delays while in the queue raise the risk of new ice accumulation and provide an incentive to go back to repeat the de-icing. Leaving the queue so that the plane can be de-iced again will cause additional delays; the aircraft will have to re-enter the takeoff queue again. The organization of activities, notably the timing of de-icing and impact of a re-de-icing on location in the queue, can create conflicts. For individual cases, practitioners resolve these conflicts through action, that is, by deciding to return to the de-icing station or remaining in line to takeoff.

Conventional human factors task analyses do not pick up such tradeoffs – task analyses operate at too microscopic a grain of analysis, and how to resolve these conflicts is rarely part of formal job descriptions. The strategic dilemmas may not arise as an explicit conscious decision by an individual so that knowledge acquisition sessions with an expert may not reveal its presence.

To evaluate the behavior of the practitioners involved in an incident, it is important to elucidate the relevant goals, the interactions among these goals, and the factors that influenced criterion setting on how to make tradeoffs in particular situations. The role of these factors is often missed in evaluations of the behavior of practitioners. As a result, it is easy for organizations to produce what appear to be solutions that in fact exacerbate conflict between goals rather than help practitioners handle goal conflicts in context. In part, this occurs because it is difficult for many organizations (particularly in regulated industries) to admit that goal conflicts and tradeoff decisions arise. However distasteful to admit or whatever public relations problems it creates, denying the existence of goal interactions does not make such conflicts disappear and is likely to make them even tougher to handle when they are relevant to a particular situation. As Feynman remarked regarding the Challenger disaster, "For a successful technology, reality must take precedence over public relations, for nature cannot be fooled" (Rogers et al., 1986, Appendix F, p. 5). The difference is that, in human-machine systems, one can sweep the consequences of attempting to fool nature under the rug by labeling the outcome as the consequence of "human error."

CONCLUSION

When you feel you have to ask, "How could people have missed …? or how could they not have known?" remind yourself to go back and trace knowledge, mindset and goal conflicts as the situation evolved. Try to understand how knowledge was brought to bear in context by people trying to solve an operational problem. Try to see how the cues and indications about their world, and how they evolved and came in over time, influenced what people understood their situation to be at the time and where they reasonably decided they should direct their attention next. Try to grasp how multiple interacting goals, many of them conflicting, some expressed more subtly then others, influenced people's trade-offs, preferences and priorities. Taken together, this is the recipe for escaping the hindsight bias. This technique, or approach, is laid out in much more detail in *The Field Guide to Understanding Human Error* (Dekker, 2002).

Also, when your organization considers making changes to the system, the three topics of this section can help you map out the potential cognitive system consequences or reverberations of change. How will changes to your system affect people's ability to smoothly, fluidly move their attentional resources as operational situations develop around them? How will changes to your system alter the way knowledge is packaged, delivered, transmitted, stored, and organized across the various parts of your system (human and technological)? And how, by extension, does this impact people's ability to bring knowledge to bear in actual settings where it is needed? Finally, how will changes to your system influence the weighting and prominence of certain goals over others, in turn shifting operational people's trade-off points or sensitivity to particular strategic directions? These are the systemic reverberations that will influence your system's ability to create success and forestall failure.

PART IV
HOW DESIGN CAN INDUCE ERROR

❖

This part of the book describes several classic deficiencies in computerized devices and how these negatively influence practitioner cognition and collaboration. Characteristics of computerized devices that shape cognition and collaboration in ways that increase the potential for error are one type of problem that can contribute to incidents. The presence of these characteristics, in effect, represents a failure of design in terms of operability (i.e., a kind of design "error"). We will show why these device characteristics are deficiencies, and we will show how the failure to design for effective human-computer cooperation increases the risk of bad outcomes.

The first chapter of this part deals with what we've called clumsy automation and one of its results: automation surprises. Automation surprises are situations where crews are surprised by actions taken (or not taken) by the automatic system, and we have examples from both advanced flight decks and the operating theater. Automation surprises begin with miscommunication and misassessments between the automation and users, which lead to a gap between the user's understanding of what the automated systems are set up to do, what they are doing, and what they are going to do. The initial trigger for such a mismatch can arise from several sources like erroneous inputs such as mode errors or indirect mode changes where the system autonomously changes its status and behavior based on its interpretation of pilot inputs, its internal logic or sensed environmental conditions. The gap results in people being surprised when the system behavior does not match their expectations. This can lead to detrimental consequences in safety-critical settings.

The second chapter of this part attempts to map in more detail how computer-based artifacts shape people's cognition and collaboration. How a problem is represented influences the cognitive work needed to solve that problem, which either improves or degrades performance. The chapter traces how technology impacts people's cognition; how cognition impacts people's behavior in operational setting, and how such behavior can contribute to an incident's evolution. Computers have a huge impact here. A fundamental property, after all, of the computer as a medium for representation is freedom from the physical constraints acting on the real-world objects/systems that the representation (i.e.

the things on a computer screen) refers to. Such virtuality carries a number of penalties, including people getting lost in display page architectures, and the hiding of interesting changes, events and system behaviors.

The third chapter is dedicated to one of the most vexing problems in human-computer interaction: mode errors. These occur when an intention is executed in a way appropriate for one mode when, in fact, the system is in a different mode. The complexity of modes, interactions across modes, and indirect mode changes create new paths for errors and failures. No longer are modes only selected and activated through deliberate explicit actions. Rather, modes can change as a side effect of other practitioner actions or inputs depending on the system status at the time. The active mode that results may be inappropriate for the context, but detection and recovery can be very difficult in part due to long time-constant feedback loops.

This chapter also discusses one consequence of mode confusion: the subsequent going-sour scenario that seems basic to a number of incidents in complex, highly computerized systems. In this scenario, minor disturbances, misactions, miscommunications and miscoordinations collectively manage a system into hazard despite multiple opportunities (certainly in hindsight) to detect that the situation is headed for a negative outcome.

The final chapter of this part considers the research results on how people adapt to new technology. We identify several types of practitioner adaptation to the impact of new information technology. In system tailoring, practitioners adapt the device and context of activity to preserve existing strategies used to carry out tasks (e.g., adaptation focuses on the set-up of the device, device configuration, how the device is situated in the larger context). In task tailoring, practitioners adapt their strategies, especially cognitive and collaborative strategies, for carrying out tasks to accommodate constraints imposed by the new technology. User adaptations (or user tailoring) can sometimes be brittle – working well in a narrow range of routine situations, but quite vulnerable when operational conditions push the user off a familiar pathway. The part finishes with a discussion of use-centered design.

9
CLUMSY USE OF TECHNOLOGY

TECHNOLOGY CHANGE TRANSFORMS OPERATIONAL AND COGNITIVE SYSTEMS

There are several possible motivations for studying an operational system in relation to the potential for error and failure. The occurrence of an accident or a near miss is a typical trigger for an investigation. Cumulated evidence from incident data bases may also provide a trigger to investigate "human error."

Another important trigger for examining the potential for system breakdown is at points of major technology change. Technology change is an intervention into an ongoing field of activity (Winograd and Flores, 1987; Flores, Graves, Hartfield, and Winograd, 1988; Carroll, Kellogg, and Rosson, 1991). When developing and introducing new technology, one should realize that the technology change represents new ways of doing things; it does not preserve the old ways with the simple substitution of one medium for another (e.g., paper for computer-based).

Technological change is, in general, transforming the workplace through the introduction and spread of new computer-based systems (Woods and Dekker, 2000). First, ubiquitous computerization has tremendously advanced our ability to collect, transmit, and transform data. In all areas of human endeavor, we are bombarded with computer-processed data, especially when anomalies occur. But our ability to digest and interpret data has failed to keep pace with our abilities to generate and manipulate greater and greater amounts of data. Thus, we are plagued by data overload.

Second, user interface technology has allowed us to concentrate this expanding field of data into one physical platform, typically a single visual display unit (VDU). Users are provided with increased degrees of flexibility for data handling and presentation in the computer interface through window management and different ways to display data. The technology provides the capability to generate tremendous networks of computer displays as a kind of virtual perceptual field viewable through the narrow aperture of the VDU. These changes affect the cognitive demands and processes associated with extracting meaning from large fields of data.

Third, heuristic and algorithmic technologies expand the range of subtasks and cognitive activities that can be automated. Automated resources can, in principle, offload practitioner tasks. Computerized systems can be developed that assess or diagnose the situation at hand, alerting practitioners to various concerns and advising practitioners on possible responses. These "intelligent" machines create joint cognitive systems that distribute cognitive work across multiple agents. Automated and intelligent agents change the composition of the team and shift the human's role within that cooperative ensemble (see Hutchins, 1995a, 1995b for treatments of how cognitive work in distributed across agents).

One can guard against the tendency to see automation in itself as a cure for "human error" by remembering this syllogism (Woods and Hollnagel, 2006, p. 176):

○ All cognitive systems are finite (people, machines, or combinations). All finite cognitive systems in uncertain changing situations are fallible. Therefore, machine cognitive systems (and joint systems across people and machines) are fallible.

We usually speak of the fallibility of machine "intelligence" in terms of brittleness – how machine performance breaks down quickly on problems outside its area of competence (cf., Roth et al., 1987; Guerlain et al., 1996). The question, then, is not the universal fallibility or finite resources of systems, but rather the development of strategies that handle the fundamental tradeoffs produced by the need to act in a finite, dynamic, conflicted, and uncertain world.

Fourth, computerization and automation integrate or couple more closely together different parts of the system. Increasing the coupling within a system has many effects on the kinds of cognitive demands practitioners face. For example, with higher coupling, actions produce more side effects. Fault diagnosis becomes more difficult as a fault is more likely to produce a cascade of disturbances that spreads throughout the monitored process. Increased coupling creates more opportunities for situations to arise with conflicts between different goals.

Technology change creates the potential for new kinds of error and system breakdown as well as changing the potential for previous kinds of trouble. Take the classic simple example of the transition from an analog alarm clock to a digital one. With the former, errors are of imprecision – a few minutes off one way or another. With the advent of the latter, precision increases, but it is now possible for order-of-magnitude errors where the alarm is set to sound exactly 12 hours off (i.e., by confusing PM and AM modes). "Design needs to occur with the possibility of error in mind" (Lewis and Norman, 1986). Analysis of the potential for system breakdown should be a part of the development process for all technology changes.

This point should not be interpreted as part of a go/no go decision about new technology. It is not the technology itself that creates the problem; rather it is how the technological possibilities are utilized vis à vis the constraints and needs of the operational system. One illustration of the complex reverberations of technology change comes from this internal reflection on the impact of the new computer technology used in NASA's new mission control in 1996 (personal communication, NASA Johnson Space Center, 1996):

We have much more flexibility in our how our displays look and in the layout of the displays on the screens. We also have the added capabilities that allow the automation of the monitoring of telemetry. But when something has advantages, it usually has disadvantages, and the new consoles are no exception. First, there is too much flexibility, so much stuff to play with that it can get to the point where adjusting stuff on the console distracts from keeping up with operations. The configuration of the displays, the various supporting applications, the ability to 'channel surf' on the TV, all lead to a lack of attention to operations. I have seen teams miss events or not hear calls on the loops because of being preoccupied with the console. I have also witnessed that when a particular application doesn't work, operations were missed due to trying to troubleshoot the problem. And this was an application that was not critical to the operations in progress. ... Second, there's too much reliance on automation, mainly the Telemetry Monitor program. I'm concerned that it is becoming the prime (and sometimes sole) method for following operations. When the crew is taught to fly the arm, they are trained to use all sources of feedback, the D&C panel, the window views, multiple camera views, and the spec. When we 'fly' the console, we must do the same. This point was made very evident during a recent sim when Telemetry Monitor wasn't functioning. It took the team awhile to notice that it wasn't working because they weren't cross checking and then once they realized it they had some difficulty monitoring operations. If this were to happen in flight it could, at a minimum, be embarrassing, and, at a maximum, lead to an incorrect failure diagnosis or missing a failure or worse – such as a loss of a payload. The solution to this problem is simple. We need to exercise judgment to prioritize tending to the console vs. following operations. If something is required right now, fix it or work around it. If it's not required and other things are going on, let it wait.

This commentary by someone involved in coping with the operational effects of technology change captures a pattern found in research on the effects new levels of automation. New levels of automation transform operational systems. People have new roles that require new knowledge, new attentional demands, and new forms of judgment. Unfortunately, the NASA manager had one thing wrong – developing training to support the "judgment to prioritize" between doing the job versus tending to the interface has not proven to be a simple matter.

PATTERNS IN THE CLUMSY USE OF COMPUTER TECHNOLOGY

We usually focus on the perceived benefits of new automated or computerized devices and technological aids. Our fascination with the possibilities afforded by technology in general often obscures the fact that new computerized and automated devices also create new burdens and complexities for the individuals and teams of practitioners responsible for operating, troubleshooting, and managing high-consequence systems. The demands may involve new or changed tasks such as device setup and initialization, configuration control, or operating sequences. Cognitive demands change as well, creating new interface management tasks, new attentional demands, the need to track automated device state and performance, new communication or coordination tasks, and new knowledge requirements. These demands represent new levels and types of operator workload.

The dynamics of these new demands are an important factor because in complex systems human activity ebbs and flows, with periods of lower activity and more self-paced tasks interspersed with busy, high-tempo, externally paced operations where task performance is more critical. Technology is often designed to shift workload or tasks from the human to the machine. But the critical design feature for well integrated cooperative cognitive work between the automation and the human is not the overall or time-averaged task workload. Rather, it is how the technology impacts on low-workload and high-workload periods, and especially how it impacts on the practitioner's ability to manage workload that makes the critical difference between clumsy and skillful use of the technological possibilities.

A syndrome, which Wiener (1989) has termed "clumsy automation," is one example of technology change that in practice imposes new burdens as well as some of the expected benefits. Clumsy automation is a form of poor coordination between the human and machine in the control of dynamic processes where the benefits of the new technology accrue during workload troughs, and the costs or burdens imposed by the technology occur during periods of peak workload, high-criticality, or high-tempo operations. Despite the fact that these systems are often justified on the grounds that they would help offload work from harried practitioners, we find that they in fact create new additional tasks, force the user to adopt new cognitive strategies, require more knowledge and more communication at the very times when the practitioners are most in need of true assistance. This creates opportunities for new kinds of human error and new paths to system breakdown that did not exist in simpler systems.

To illustrate these new types of workload and their impact on practitioner cognition and collaboration let us examine two series of studies, one looking at pilot interaction with cockpit automation, and the other looking at physician interaction with new information technology in the operating room. Both series of studies found that the benefits associated with the new technology accrue during workload troughs, and the costs associated with the technology occur during high-criticality, or high-tempo operations.

CLUMSY AUTOMATION ON THE FLIGHT DECK

Results indicate that one example of clumsy automation can be seen in the interaction between pilots and flight management computers (FMCs) in commercial aviation. Under low-tempo operations pilots communicate instructions to the FMCs which then "fly" the aircraft. Communication between pilot and FMC occurs through a multi-function display and keyboard. Instructing the computers consists of a relatively effortful process involving a variety of keystrokes on potentially several different display pages and a variety of cognitive activities such as recalling the proper syntax or where data is located in the virtual display page architecture. Pilots speak of this activity as "programming the FMC."

Cockpit automation is flexible also in the sense that it provides many functions and options for carrying out a given flight task under different circumstances. For example, the FMC provides at least five different mechanisms at different levels of automation for changing altitude. This customizability is construed normally as a benefit that allows the pilot to select the mode or option best suited to a particular flight situation (e.g., time and speed constraints). However, it also creates demands for new knowledge and new judgments. For example, pilots must know about the functions of the different modes, how to coordinate which mode to use when, and how to "bumplessly" switch from one mode or level of automation to another. In other words, the supervisor of automated resources must not only know something about how the system works, but also know how to work the system. Monitoring and attentional demands are also created as the pilots must keep track of which mode is active and how each active or armed mode is set up to fly the aircraft.

In a series of studies on pilot interaction with this suite of automation and computer systems, the data revealed aspects of cockpit automation that were strong but sometimes silent and difficult to direct when time is short. The data showed how pilots face new challenges imposed by the tools that are supposed to serve them and provide "added functionality." For example, the data indicated that it was relatively easy for pilots to lose track of the automated systems' behavior during high-tempo and highly dynamic situations. Pilots would miss mode changes that occurred without direct pilot intervention during the transitions between phases of flight or during the high-workload descent and approach phases in busy airspace. These difficulties with mode awareness reduced pilots' ability to stay ahead of the aircraft.

Pilots develop strategies to cope with the clumsiness and complexities of many modern cockpit systems. For example, data indicate that pilots tend to become proficient or maintain their proficiency on a subset of modes or options. As a result, they try to manage the system within these stereotypical responses or paths, underutilizing system functionality. The results also showed that pilots tended to abandon the flexible but complex modes of automation and switch to less automated, more direct means of flight control, when the pace of operations increased (e.g., in crowded terminal areas where the frequency of changes in instructions from air traffic control increase). Note that pilots speak of this tactic as "escaping" from the automation.

From this and other research, the user's perspective on the current generation of automated systems is best expressed by the questions they pose in describing incidents (Wiener, 1989):

- What is it doing now?
- What will do next?
- How did I get into this mode?
- Why did it do this?
- Stop interrupting me while I am busy.
- I know there is some way to get it to do what I want.
- How do I stop this machine from doing this?
- Unless you stare at it, changes can creep in.

These questions and statements illustrate why one observer of human-computer interaction defined the term agent as "A computer program whose user interface is so obscure that the user must think of it as a quirky, but powerful, person" (Lanir, 1995,

p. 68, as quoted in Woods and Hollnagel, 2006, p. 120). In other words, the current generation of cockpit automation contains several classic human-computer cooperation problems, e.g., an opaque interface.

Questions and statements like these point to automation surprises, that is, situations where crews are surprised by actions taken (or not taken) by the automated system. Automation surprises begin with miscommunication and misassessments between the automation and users which lead to a gap between the user's understanding of what the automated systems are set up to do, what they are doing, and what they are going to do. The initial trigger for such a mismatch can arise from several sources, for example, erroneous inputs such as mode errors or indirect mode changes where the system autonomously changes states based on its interpretation of pilot inputs, its internal logic and sensed environmental conditions. The gap results from poor feedback about automation activities and incomplete mental models of how it functions. Later, the crew is surprised when the aircraft's behavior does not match the crew's expectations. This is where questions like, "Why won't it do what I want?" "How did I get into this mode?" arise.

When the crew is surprised, they have detected the gap between expected and actual aircraft behavior, and they can begin to respond to or recover from the situation. The problem is that this detection generally occurs when the aircraft behaves in an unexpected manner – flying past the top of descent point without initiating the descent, or flying through a target altitude without leveling off. In other words, the design of the pilot-automation interface restricts the crews' ability to detect and recover from the miscoordination. If the detection of a problem is based on actual aircraft behavior, it may not leave a sufficient recovery interval before an undesired result occurs (low error-tolerance). Unfortunately, there have been accidents where the misunderstanding persisted too long to avoid disaster.

CLUMSY AUTOMATION IN THE OPERATING ROOM: 1 – CENTRALIZING DATA DISPLAY

Another study, this time in the context of operating room information systems, reveals some other ways that new technology creates unintended complexities and provokes practitioner coping strategies (Cook and Woods, 1996b). In this case a new operating room patient monitoring system was studied in the context of cardiac anesthesia. This and other similar systems integrate what was previously a set of individual devices, each of which displayed and controlled a single sensor system, into a single CRT display with multiple windows and a large space of menu-based options for maneuvering in the space of possible displays, options, and special features. The study consisted of observing how the physicians learned to use the new technology as it entered the workplace.

By integrating a diverse set of data and patient monitoring functions into one computer-based information system, designers could offer users a great deal of customizability and options for the display of data. Several different windows could be called depending on how the users preferred to see the data. However, these flexibilities all created the need for the physician to interact with the information system – the physicians had to direct attention to the display and menu system and recall knowledge about the system.

Furthermore, the computer keyhole created new interface management tasks by forcing serial access to highly inter-related data and by creating the need to periodically declutter displays to avoid obscuring data channels that should be monitored for possible new events.

The problem occurs because of a fundamental relationship: the greater the trouble in the underlying system or the higher the tempo of operations, the greater the information processing activities required to cope with the trouble or pace of activities. For example, demands for monitoring, attentional control, information, and communication among team members (including human-machine communication) all tend to go up with the tempo and criticality of operations. This means that the burden of interacting with the display system tends to be concentrated at the very times when the practitioner can least afford new tasks, new memory demands, or diversions of his or her attention away from patient state to the interface per se.

The physicians tailored both the system and their own cognitive strategies to cope with this bottleneck. In particular, they were observed to constrain the display of data into a fixed spatially dedicated default organization rather than exploit device flexibility. They forced scheduling of device interaction to low criticality self-paced periods to try to minimize any need for interaction at high workload periods. They developed stereotypical routines to avoid getting lost in the network of display possibilities and complex menu structures.

CLUMSY AUTOMATION IN THE OPERATING ROOM: 2 – REDUCING THE ABILITY FOR RECOVERY FROM ERROR OR FAILURE

This investigation started with a series of critical incidents involving physician interaction with an automatic infusion device during cardiac surgery. The infusion controller was a newly introduced computer-based device used to control the flow of blood pressure and heart rate medications to patients during heart surgery. Each incident involved delivery of a drug to the patient when the device was supposed to be off or halted. Detailed debriefing of participants suggested that, under certain circumstances, the device would deliver drug (sometimes at a very high rate) with little or no evidence to the user that the infusion was occurring. A series of investigations were done including observation of device use in context to identify:

o   characteristics of the device which make its operation difficult to observe and error prone and,
o   characteristics of the context of cardiac anesthesiology which interact with the device characteristics to provide opportunities for unplanned delivery of drug (Cook et al., 1992; Moll van Charante et al., 1993).

In cardiac surgery, the anesthesiologist monitors the patient's physiological status (e.g., blood pressure, heart rate) and administers potent vasoactive drugs to control these parameters to desired levels based on patient baselines, disease type, and stage of cardiac surgery. The vasoactive drugs are administered as continuous infusion drips mixed with intravenous fluids. The device in question is one type of automatic infusion controller that regulates the rate of flow. The user enters a target in terms of drops per minute, the device counts drops that form in a drip chamber, compares this to the target, and adjusts flow. If the device is unable to regulate flow or detects one of several different device conditions, it is programmed to cease operation and emit an audible alarm and warning message. The interface controls consist of three multi-function buttons and a small LCD panel which displays target rate and messages. In clinical use in cardiac surgery up to six devices may be set up with different drugs that may be needed during the case.

The external indicators of the device's state provide poor feedback and make it difficult for physicians to assess or track device behavior and activities. For example, the physician users were unaware of various controller behavioral characteristics such as overshoot at slow target rates, "seek" behavior, and erratic control during patient transport. Alarms were remarkably common during device operation. The variety of different messages were ambiguous – several different alarm messages can be displayed for the same underlying problem; the different messages depend on operating modes of the device which are not indicated to the user. Given the lack of visible feedback, when alarms recurred or a sequence occurred, it was very difficult for the physician to determine whether the device had delivered any drug in the intervening period.

The most intense periods of device use also were those time periods of highest cognitive load and task criticality for the physicians, that is, the time period of coming off cardio-pulmonary bypass. It is precisely during these periods of high workload that the automated devices are supposed to provide assistance (less user workload through more precise flows, smoother switching between drip rates, and so on). However, this was also the period where the largest number of alarms occurred and where device troubleshooting was most onerous.

Interestingly, users seemed quite aware of the potential for error and difficulties associated with device setup which could result in the device not working as intended when needed. They sought to protect themselves from these troubles in various ways, although the strategies were largely ineffective.

In the incidents, misassemblies or device problems led to inadvertent drug deliveries. The lack of visible feedback led physicians to think that the device was not delivering drug and was not the source of the observed changes in patient physiology. Large amounts of vasoactive drugs were delivered to brittle cardiovascular systems, and the physicians were unable to detect that the infusion devices were the source of the changes. Luckily in all of the cases, the physicians responded appropriately to the physiological changes with other therapies and avoided any adverse patient outcomes. Only later did the physicians realize that the infusion device was the source of the physiological changes. In other words, these were cases of automation surprise.

The investigations revealed that various device characteristics led to an increased potential for misassessments of device state and behavior. These characteristics played a role in the incidents because they impaired the physician's ability to detect and recover from unintended drug administrations. Because of these effects, the relevant characteristics of the device can be seen as deficiencies from a usability point of view; the device design is "in error."

The results of this series of studies directly linked, for the same device and context, characteristics of computerized devices to increased potential for erroneous actions and impaired ability to detect and recover from errors or failures. Furthermore, the studies directly linked the increased potential for erroneous setup and the decreased ability to detect errors as important contributors to critical incidents.

THE IMPACT OF CLUMSY AUTOMATION ON COGNITION AND COLLABORATION

There are some important patterns in the results from the above studies and others like them. One is that characteristics of computer-based devices and systems affect the potential for different kinds of erroneous actions and assessments. Characteristics of computer-based devices that influence cognition and behavior in ways that increase the potential for erroneous actions and assessments can be considered flaws in the joint human-computer cognitive system that can create operational problems for people at the sharp end.

A second pattern is that the computer medium shapes the constraints for design. In pursuit of the putative benefits of automation, user customizability, and interface configurability, it is easy for designers to unintentionally create a thicket of modes and options, to create a mask of apparent simplicity overtop of underlying device or interface complexity, to create a large network of displays hidden behind a narrow keyhole.

A result that occurred in all the above studies is that practitioners actively adapted or tailored the information technology provided for them to the immediate tasks at hand in a locally pragmatic way, usually in ways not anticipated by the designers of the information technology. Tools are shaped by their users.

New technology introduced for putative benefits in terms of human performance in fact introduced new demands and complexities into already highly demanding fields of practice. Practitioners developed and used a variety of strategies to cope with these new complexities. Because practitioners are responsible agents in the domain, they work to insulate the larger system from device deficiencies and peculiarities of the technology. This occurs, in part, because practitioners inevitably are held accountable for failure to correctly operate equipment, diagnose faults, or respond to anomalies even if the device setup, operation, and performance are ill-suited to the demands of the work environment.

In all of these studies practitioners tailored their strategies and behavior to avoid problems and to defend against device idiosyncrasies. However, the results also show how these adaptations were only partly successful. The adaptations could be effective, or only locally adaptive, in other words, brittle to various degrees (i.e., useful in narrow contexts, but problematic in others).

An underlying contributor to the above problems in human-automation coordination is the escalation principle. There is a fundamental relationship where the greater the trouble in the underlying process or the higher the tempo of operations, the greater the information processing activities required to cope with the trouble or pace of activities. For example, demands for monitoring, attentional control, information, and communication among team members (including human-machine communication) all tend to go up with the unusualness (situations at or beyond margins of normality or beyond textbook situations), tempo and criticality of situations. If there are workload or other burdens associated with using a computer interface or with interacting with an autonomous or intelligent machine agent, these burdens tend to be concentrated at the very times when the practitioner can least afford new tasks, new memory demands, or diversions of his or her attention away from the job at hand to the interface per se. This is the essential trap of clumsy automation.

Finally, it would be easy to label the problems noted above as simply "human-computer interaction deficiencies." In some sense they are exactly that. But the label "human-computer interaction" (HCI) carries with it many different assumptions about the nature of the relationship between people and technology. The examples above illustrate deficiencies that go beyond the concepts typically associated with the label computer interface in several ways.

First, all of these devices more or less meet guidelines and common practices for human-computer interaction defined as simply making the needed data nominally available, legible, and accessible. The characteristics of the above systems are problems because of the way that they shape practitioner cognition and collaboration in their field of activity. These are not deficiencies in an absolute sense; whether or not they are flaws depends on the context of use.

Second, the problems noted above cannot be seen without understanding device use in context. Context-free evaluations are unlikely to uncover the important problems, determine why they are important, and identify criteria that more successful systems should meet.

Third, the label HCI easily conjures up the assumption of a single individual alone, rapt in thought, but seeing and acting through the medium of a computerized device. The cases above and the examples throughout this volume reveal that failures and successes involve a system of people, machine cognitive agents, and machine artifacts embedded in context. Thus, it is important to see that the deficiencies, in some sense, are not in the computer-based device itself. Yes, one can point to specific aspects of devices that contribute to problems (e.g., multiple modes, specific opaque displays, or virtual workspaces that complicate knowing where to look next), but the proper unit of analysis is not the device or the human. Rather, the proper unit of analysis is the distributed system that accomplishes cognitive work – characteristics of artifacts are deficient because of how they shape cognition and collaboration among a distributed set of agents. Clumsiness is not really in the technology. Clumsiness arises in how the technology is used relative to the context of demands, resources, agents, and other tools.

Today, most new developments are justified in part based on their presumed impact on human performance. Designers believe:

o automating tasks will free up cognitive resources,
o automated agents will offload tasks,
o automated monitors will diagnose and alert users to trouble,
o flexible data handling and presentation will allow users to tailor data to the situation,
o integrating diverse data onto one screen will aid situation assessment.

The result is supposed to be reduced workload, enhanced productivity, and fewer errors.

In contrast, studies of the impact of new technology on users, like those above, tell a very different story. We see an epidemic of clumsy use of technology, creating new burdens for already beleaguered practitioners, often at the busiest or most critical times. Data shows that:

- instead of freeing up resources, clumsy use of technology creates new kinds of cognitive work,
- instead of offloading tasks, autonomous, but silent machine agents create the need for team play, coordination and intent communication with people, demands which are difficult for automated systems to meet,
- instead of focusing user attention, clumsy use of technology diverts attention away from the job to the interface,
- instead of aiding users, generic flexibilities create new demands and error types,
- instead of reducing human error, clumsy technology contains "classic" flaws from a human-computer cooperation point of view; these create the potential for predictable kinds of erroneous actions and assessments by users,
- instead of reducing knowledge requirements, clumsy technology demands new knowledge and more difficult judgments.

10

HOW COMPUTER-BASED ARTIFACTS SHAPE COGNITION AND COLLABORATION

A MAP: THE IMPACT FLOW DIAGRAM

Figure 10.1 provides an overall map of the process by which the clumsy use of new computer technology affects the cognition and behavior of people embedded in an operational system, creating the potential for problems which could contribute to incidents or accidents. The figure is a schematic of the results of research on the relationship of computer technology, cognition, practitioner behavior, and system failure. We will refer to it as the Impact Flow Diagram because it maps:

- how technology impacts cognition in context,
- how cognition impacts behavior in operational contexts, and
- how behavior can contribute to incident evolution.

Artifacts shape cognition and collaboration. This is one of the fundamental findings of Cognitive Science. How a problem is represented influences the cognitive work needed to solve that problem, either improving or degrading performance – the representation effect (excellent demonstrations and treatments of the representation effect can be found in Norman (1993), Zhang and Norman (1994) and Zhang (1997)). Don Norman (1993) has expressed this effect in more colloquial terms, "things can make us smart" and "things can make us dumb."

The representation effect means that we should think of computer-based artifacts in terms of how they represent the underlying process for some observer in some goal and task context (Woods, 1996). Characteristics of devices shape cognition and collaboration depending on the context of activities, demands, and goals in the particular field of activity.

When "things make us dumb," technology impairs rather than supports cognitive work. When computerized devices:

- make things invisible, especially hiding "interesting" events, changes, and anomalies,

- proliferate modes,
- force serial access to highly related data,
- proliferate windows and displays in virtual-data space behind a narrow keyhole,
- contain complex and arbitrary sequences of operations, modes, and mappings,
- add new interface-management tasks that tend to congregate at high-criticality and high-tempo periods of the task,
- suppress cues about the activities of other team members both machine and human,

they misrepresent the underlying process and undermine the cognitive activities of practitioners.

Such "clumsy" systems do this because they (Figure 10.1):

- increase demands on user memory,
- complicate situation assessment,
- undermine attentional control skills (where to focus when),
- add workload at high-criticality high-tempo periods,
- constrain the users' ability to develop effective workload management strategies,
- impair the development of accurate mental models of the function of the device and the underlying processes,
- decrease knowledge calibration (i.e., mislead users into thinking that their models are more accurate than they actually are),
- undermine coordination across multiple agents.

Only by examining how the computerized system represents the behavior of the underlying process in question can one see these representational flaws. How a display or interface represents is bound to the context of the demands of underlying process and the goals and tasks of practitioners who manage that process.

The impact of computer-based systems on cognition and collaboration is important because it influences how practitioners behave in various situations that can arise in that field of activity. Clumsy use of technology ultimately impacts the behavior of practitioners producing:

- increased potential for different kinds of erroneous actions and erroneous assessments of process state (e.g., mode errors),
- impaired ability to detect and recover from failures, erroneous actions, or assessments,
- creates the need for practitioners to tailor their behavior and the device to make it into a more usable tool, tailoring that may create vulnerabilities to human-machine system breakdowns in special circumstances,
- increased risk of falling behind in incident evolution (loss of situation awareness and other breakdowns in mindset),
- automation surprises or other breakdowns in coordination across multiple agents,
- decreased learning opportunities.

HOW COMPUTER-BASED ARTIFACTS SHAPE COGNITION AND COLLABORATION

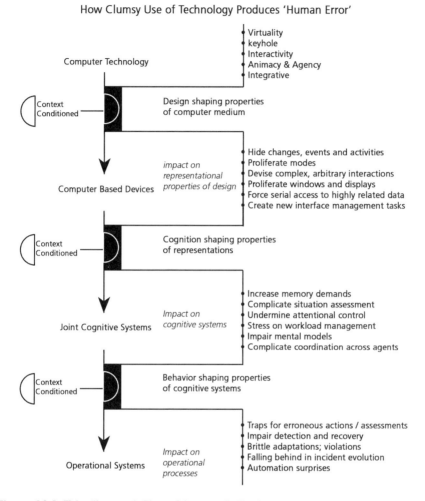

Figure 10.1 This "Impact Flow Diagram" illustrates the relationship between the design-shaping properties of the computer as a medium, the cognition-shaping properties of representations in the computer medium, and the behavior-shaping properties of cognitive systems. The impact flow relationships define a failure path for the clumsy use of technology

The Impact Flow Diagram shows how the clumsy use of technological possibilities shapes the cognition and behavior of the people embedded in the operational system in predictable ways. There are design-shaping properties of the computer medium that make it easy for designers to create devices with typical flaws in human-computer cooperation. These characteristics are flaws because they create new cognitive demands and increase the stress on other cognitive activities. Behavior-shaping properties of cognitive systems link these effects to different kinds of operational consequences. As a result, these design deficiencies become problems that can contribute to incidents and accidents, if other potentiating factors are present. When investigators go behind the label human error to find the Second Story of an incident, they often find issues in the design of the technological systems intended by developers as user "support" systems. The irony is that the very characteristics of computer-based devices that have been shown to complicate practitioners' cognitive activities and contribute to failures are generally justified and marketed on the grounds that they reduce human workload and improve human performance.

"CLASSIC" FLAWS IN HUMAN-COMPUTER COOPERATION: DESIGNER "ERROR"

This section provides a brief discussion of how human computer cooperation flaws can be thought of as design failures that arise from properties of the computer as a medium for representation and from factors that influence how new technology is deployed. These characteristics of computer-based devices are "flaws" because of how they predictably burdens practitioner cognition and collaboration.

A complete treatment of "flaws" in human-computer cooperation would require a volume in its own right. A broad treatment also would evoke another need – aiding the designer of human-computer systems in the development of systems that improve operational performance in a particular setting. This section is not intended as a designer's guide, but simply to help the reader see how technological change influences people and can result in new types of difficulties and failure paths (the relationships in the Impact Flow Diagram). First, we will provide an overview of some of the typical flaws and how they arise. Second, we will provide a comprehensive treatment of one of these flaws – mode error, including some potential countermeasures. Finally, we will use the mode error case to illustrate the relationships captured in the Impact Flow Diagram.

PENALTIES OF VIRTUALITY

> Every parameter you can control, you must control. (W. Carlos' First Law of Digital Synthesized Music, 1992)

A fundamental property of the computer as a medium for representation is freedom from the physical constraints acting on the referent real world objects/systems (Woods, 1996, derived this property of computers from Hochberg's analysis of perception mediated by different media, 1986, pp. 222–223). In many media (e.g., cinema), the structure and constraints operating in the physical world will ensure that much of the appropriate "information" about relationships in the referent domain is preserved in the representation. On the other hand in the computer medium, the designer of computer displays of data must do all of the work to constrain or link attributes and behaviors of the representation to the attributes and behaviors of the referent domain.

This property means that sets of displays of data in the computer medium can be thought of as a virtual perceptual field. It is a perceivable set of stimuli, but it differs from a natural perceptual field and other media for representation because there is nothing inherent in the computer medium that constrains the relationship between things represented and their representation. This freedom from the physical constraints acting on the referent real-world objects is a double-edged sword in human-computer cooperation, providing at the same time the potential for very poor representations (things that make us dumb) and the potential for radically new and more effective representations (things that make us smart).

The computer medium allows designers to combine multiple features, options, and functions onto a single physical platform. The same physical device can be designed to operate in many different contexts, niches, and markets simply by taking the union of all the features, options, and functions that are needed in any of these settings. In a sense, the computer medium allows one to create multiple virtual devices concatenated onto a single physical device. After all, the computer medium is multi-function – software can make the same keys do different things in different combinations or modes, or provide soft keys, or add new options to a menu structure; the visual display unit (VDU) allows one to add new displays which can be selected if needed to appear on the same physical viewport. It is the 'ne plus ultra' in modular media.

But this means that a practitioner cannot have the device in one context without also importing part of the complexity from all of the other contexts. Concatenating multiple virtual devices on a single platform forces practitioners concerned with only a single niche to deal with the complexity of all the other niches as well. This is in contradiction to what people are observed to do to cope with complexity – people divide up a domain to segregate the complexity in ways that are meaningful, that is, into a series of local contexts. Furthermore, it is a fundamental research result that human cognition and behavior are conditioned to the context in which they occur.

The virtuality of computer-based information technology allows designers to develop new subsystems or devices with the appearance of simplicity by integrating diverse data, capabilities, and devices into a single multi-function VDU display and interface. But to do this pushes the designer to proliferate modes, to proliferate displays hidden behind the narrow viewport, to assign multiple functions to controls, to devise complex and arbitrary sequences of operation – in other words, to follow Don Norman's (1988) tongue-in-cheek advice on how to do things wrong in designing computer-based devices. Such systems appear on the surface to be simple because they lack large numbers of physical display devices and controls. However, underneath the placid surface of the VDU workstation there may be a variety of characteristics which produce cognitive burdens and operational complexities.

For example, it is easy to design systems where a few keys do many things in combination, but from the practitioners' perspective, this is very likely to create complex and arbitrary control sequences. The result can be memory burdens and fertile ground for a variety of phenotypical action errors such as omissions, repetitions, and for genotypical patterns in action errors (Cook et al., 1991b; Obradovich and Woods, 1996). Practitioners will develop coping strategies to deal with the operational and cognitive clumsiness of these complexities, for example, they create their own external memory aids (Norman, 1988). An alternative technology-centered approach provides users with a generic keypad. The ultimate in flexibility in one sense, but, from a practitioner point of view, this makes all interactions the equivalent of "programming." As a result, the costs of interacting with the device's capabilities go up, which creates bottlenecks in high-tempo periods. Practitioners cope through escape – they abandon using that device in high-tempo periods.

Getting Lost in Large Networks of Displays Hidden Behind a Narrow Keyhole

When we look at a computer based display system, what do we actually see? We directly see one display or a few displays in separate windows on one or more VDUs. All of the other displays or views of the underlying system, process, or device that could be called up onto the viewports are not directly visible. In other words, the data space is hidden behind the narrow keyhole of the VDU. The relationships between different parts of this data space are not apparent. Glancing at what is directly visible usually does not reveal the paths or mechanisms or sequence of actions required to maneuver through the data space. There is a large difference between what is directly visible and what is behind the placid surface of the VDU screen. To see another (small) portion of the data space the user must decide what they should examine, where it is in the data space, selecting it and moving it into the limited viewport (which replaces whatever portion of the data space that had been on display).

This is a description of one of the dominant characteristics of the computer as a medium for representation – the keyhole property (Woods and Watts, 1997). The computer medium exhibits the keyhole property because the size of the available viewports (the windows/VDUs available) is very small relative to the size of the artificial data space or number of data displays that potentially could be examined. As a result, the proportion of the virtual perceptual field that can be seen at the same time, that is, physically in parallel, is very very small.

To designers and technologists this property appears quite advantageous – a single physical device, the VDU, can be used to provide access in principle to any kind of view the designer, marketing department, or customer think relevant. Change or expansion is straightforward: the physical platform remains constant; one only has to modify or add more displays to the artificial space behind the physical viewports. One can respond to many different markets, niches and contexts simply by taking the union of all of the features, displays, options and functions that are needed in any of these settings and combining them on a single physical platform through software. The computer medium makes it easy to add new options to a menu structure or to add new displays which can be selected to appear on the same physical viewport if needed.

The norm is that the observer can see only one small portion of the total data field at a time or a very small number of the potentially available displays. In addition, the default tendency is to use individual pieces of data as the base unit of organization with each piece of data placed in only one location within the virtual perceptual field (one "home"). The result appears to provide users with a great deal of user configurability. They can call up, into the physical viewports available, whichever view they desire to inspect at that time.

The practitioner's view is quite different from the technologist's. The combination of a large field of raw data and a narrow keyhole creates a variety of new burdens for practitioners. Practitioners bear the burden to navigate across all of these displays in order to carry out domain tasks and to meet domain goals. One danger is that users can become "disoriented" or "lost" in the space of possible displays. There is little support for finding the right data at the right time as tasks change and activities unfold. Inter-related data becomes fragmented and spread out across different kinds of display frames forcing the practitioner into a slow serial search to collect and then integrate related data. Practitioner attention shifts to the interface (where is the desired data located in the display space?) and to interface control (how do I navigate to that location in the display space?) at the very times where their attention needs to be devoted most to their job (what is the relevant data? what does it mean about system state and needed practitioner actions?).

The keyhole property of the computer as a medium results in systems which powerfully affect the cognitive demands associated with operational tasks. Consider these three brief examples.

> One computerized control room for a nuclear power plant has evolved to include well over 16,000 possible displays that could be examined by operators. During development designers focused on individual displays and relied on a set of generic movement mechanisms and commands to support navigation across displays. Only after the control room was implemented for final testing in a high fidelity simulator did the developers realize that operators needed to focus a great deal of time and effort to find the appropriate displays for the current context. These difficulties were greatest when plant conditions were changing, the most critical times for operators to be focused on understanding and correctly responding to plant conditions.

> A device for infusing small quantities of potent medicine is used to allow mothers with high risk pregnancies to control pre-term labor while remaining at home instead of in the hospital (the patient becomes the user and the nursing staff supervises). The device interface consists of about 40 different displays nested at two levels (there are seven different "top level" displays; under each are one to seven more displays). Patient-users can only see one display at a time and must move across displays to carry out the tasks involved in control of pre-term labor and in coordinating with supervising nursing personnel. Patient-users easily become disoriented due to poor feedback about where they are in the space of possible displays and sequences of operation as well as other factors. Most commonly when lost, patient-users retreat to a default screen and attempt to begin the task over from the beginning although there are other potential consequences as well (Obradovich and Woods, 1996).

> Students were asked to learn about a topic related to their (psychology) class by studying a hypertext document of 45 cards. The hypertext document was organized either in a fixed linear sequence or in a non-linear structure where related cards were cross-referenced. No cues to structure or navigation aids were included except the variable of organization type – hierarchical or nonlinear. In the absence of any cues to structure (the non-linear condition) in this small but unfamiliar body of material, participants searched less, answered questions about the material more slowly and less accurately, and felt more disoriented.

As these cases illustrate, the keyhole property produce a sense of disorientation – losing track of where they are relative to other locations in the space of possibilities. Another set of effects are cognitive – the limited support for coordinating data across different views increases mental workload or forces users to adopt deliberative mentally demanding strategies. A third kind of impact is on performance: users may search less of the space of possibilities, miss "interesting" events, take longer to perform a task or perform with lower accuracy.

However, when practitioners perform actual work in context, getting lost represents an egregious failure to accomplish their task goals. In effect, practitioners cannot afford to get lost in poorly designed networks of displays. Observational studies of users in context reveal that practitioners actively develop strategies and modify interfaces to avoid the danger of getting lost in poorly designed networks and to minimize any navigation activities during higher tempo or more critical aspects of job performance.

This repeated observation of users actively working to avoid navigating has been noted in several different domains with several different kinds of users – anesthesiologists in the operating room, flight controllers in space mission control, financial analysts working with large extended spreadsheets. When practitioners work around navigation burdens, they are sending designers a potent message: they are saying that the interface is poorly matched to the demands of their task world.

These negative consequences of the keyhole property are not immutable. They are the result of a kind of design "error" – failing to design for across display transitions and how practitioners will coordinate a set of views as they perform meaningful tasks. There is a set of techniques to break down the keyhole inherent in computer-based display of data. Success is more than helping users travel in a space they cannot visualize. Success is creating a visible conceptual space meaningfully related to activities and constraints of a field of practice. Success lies in supporting practitioners to know where to look next in that conceptual space.

Hiding Interesting Changes, Events, and Behaviors

Typically in computer-based representations, the basic unit of display remains an individual datum usually represented as a digital value, for example oxygen tank pressure is 297 psi. Few attempts are made in the design of the representation of the monitored process to capture or highlight operationally interesting events – behaviors of the monitored process over time. This failure to develop representations that reveal change and highlight events in the monitored process has contributed to incidents where practitioners using such opaque representations miss operationally significant events.

One well-known accident where this representational deficiency contributed to the incident evolution was the *Apollo 13* mission (see Murray and Cox, 1989, for an account of mission control activities during the accident). In this accident, an explosion occurred in the oxygen portion of the cryogenics system (oxygen tank 2). One mission controller (the electrical, environmental, and communication controller or EECOM) monitoring this system was examining a screen filled with digital values (see Figure 10.2 for a recreation of this display 8 minutes before the explosion, the CSM ECS CRYO TAB display). Then he and most of the other mission controllers began seeing indications of trouble in the systems that they were responsible for monitoring. Among a host of anomalous indications, EECOM noticed that oxygen tank 2 was depressurized (about 19 psi). It took a precious 54 minutes as a variety of hypotheses were pursued before the team realized that the "command module was dying" and that an explosion in the oxygen portion of the cryogenics system was the source of the trouble. The display had hidden the critical event: two digital values, out of 54 changing digital numbers, had changed anomalously (see Figures 10.2, 10.3, and 10.4).

> So none of the three noticed the numbers for oxygen tank 2 during four particularly crucial seconds. At 55 hours, 54 minutes, and 44 seconds into the mission, the pressure stood at 996 p.s.i. – high but still within accepted limits. One second later, it peaked at 1,008 p.s.i. By 55:54:48, it had fallen to 19 p.s.i. . ... If one of them had seen the pressure continue on through the outer limits, then plunge, he would have been able to deduce that oxygen tank 2 had exploded [see Figure 10.3]. It would have been a comparatively small leap . . . to have put the whole puzzle of multiple disturbances across normally unconnected systems together. (Murray and Cox, 1989, p. 406)

Eight minutes before the explosion

```
LM1839                           CSM ECS CRYOTAB                         0613
CTE 055:46:51(    )  GET 055:46:53 (          )            SITE GDS09
------LIFE SUPPORT-----               ----PRIMARY COOLANT----
GF3571 LM CABIN P    PSIA            CF0019 ACCUM QTY    PCT    34.4
CF0001 CABIN P       PSIA     5.1    CF0016 PUMP P       PSID   45.0
CF0012 SUIT P        PSIA     4.3    SF0260 RAD IN T     °F     73.8
CF0003 SUIT ΔP     IN H2O    -1.68
CF0015 COMP ΔP     P PSID     0.30   CF0020 RAD OUT T    °F     35
CF0006 SURGE P     P PSIA      891   CF1081 EVAPINT      °F     45.6
       SURGE QTY       LB     3.67   CF0017 STEAM T      °F     64.9
    O2 TK1 CAP ΔP    PSID       21   CF0034 STEAM P      PSIA    .161
    O2 TK2 CAP ΔP    PSID       17   CF0018 EVAP OUT T   °F     44.2

CF0036 O2 MAN P      PSIA      105   SF0266 RAD VLV1/2          ONE
CF0035 O2 FLOW      LB/HR    0.181   CF0175 GLY FLO     LB/HR  215.0
CF0008 SUIT T         °F      50.5
CF0002 CABIN T        °F        65   ---SECONDARY COOLANT---
CF0005 CO2 PP        MMHG      1.5   CF0072 ACCUM QTY    PCT    36.8
----------H2O----------              CF0070 PUMP P       PSID    9.3
CF0009 WASTE          PCT     24.4   SF0262 RAD IN T     °F     76.5
       WASTE          LB      13.7   SF0263 RAD OUT T    °F     44.6
CF0010 POTABLE        PCT    104.5   CF0073 STEAM P      PSIA   .2460
       POTABLE        LB      37.6   CF0071 EVAP OUT T   °F     66.1
CF0460 URINE NOZ T    °F        70   CF0120 H2O RES      PSIA   25.0
CF0461 H2O NOZ T      °F        72   TOTAL  FC CUR       AMPS   67.50
------CRYO SUPPLY----------02-1----02-2--------H2-1------H2-2---
SC0037-38-39-40    P  PSIA     913       908       225.7       235.1
SC0032-33-30-32 QTY   PCT    77.63      01.17      73.24        73.03
SC0041-42-43-44 T     °F      -189       -192       -417         -416
                QTY   LBS    251.1      260.0      20.61        20.83
```

Figure 10.2 Eight minutes before the explosion: Reconstruction of the computer display (display CSM ECS CRYO TAB) monitored by the electrical, environmental, and communication controller (EECOM) shortly before the explosion that occurred during the *Apollo 13* mission. The pressure value for oxygen tank number two is located on the fourth line from the bottom, third column from the right. It reads 908 psi (pounds per square inch)

It was reported that the controller experienced a continuing nightmare for two weeks following the incident, in which "he looked at the screen only to see a mass of meaningless numbers" (Murray and Cox, 1989, p. 407). The display is a meaningless mass of numbers. Meaning comes from the knowledge the expert practitioner brings to bear.

Poor representations are compensated for through human adaptability and knowledge. In Norman's (1988) terminology, knowledge-in-the-head can compensate for the absence of knowledge-in-the-world. But, what is the point of the computer as a medium for the display of data if it does not reduce practitioner memory loads? And yet, in computer system after computer system we find that, despite the availability of new computational and graphic power, the end result is an increase in demands on practitioner memory. The contrast cannot be greater with studies of successful, but often technologically simple, cognitive artifacts that reveal how effective cognitive tools offload memory demands, support attentional control, and support the coordination of cognitive work across multiple agents (Hutchins, 1990, 1995a, 1995b).

Moment of the explosion

```
LM1839                           CSM ECS CRYOTAB                        0613
CTE 055:54:52 (55.914)      GET 055:54:54 (55.915)               SITE GDS09
------LIFE SUPPORT-----              ----PRIMARY COOLANT----
GF3571 LM CABIN P    PSIA            CF0019 ACCUM QTY      PCT     34.4
CF0001 CABIN P       PSIA     5.1    CF0016 PUMP P         PSID    46.0
CF0012 SUIT P        PSIA     4.3    SF0260 RAD IN T       °F      73.8
CF0003 SUIT ΔP    IN H2O   -1.72
CF0015 COMP ΔP       P PSID   0.30   CF0020 RAD OUT T      °F      35
CF0006 SURGE P     P PSIA     891    CF1081 EVAPINT        °F      46.2
       SURGE QTY      LB      3.67   CF0017 STEAM T        °F      64.9
   O2  TK1 CAP ΔP   PSID     -12     CF0034 STEAM P        PSIA    .161
   O2  TK2 CAP ΔP   PSID      105    CF0018 EVAP OUT T     °F      44.2
CF0036 O2 MAN P      PSIA     103    SF0266 RAD VLV1/2             ONE
CF0035 O2 FLOW       LB/HR    0.181  CF0175 GLY FLO        LB/HR   215.0
CF0008 SUIT T          °F     50.5
CF0002 CABIN T         °F     65     ---SECONDARY COOLANT---
CF0005 CO2 PP        MMHG     1.5    CF0072 ACCUM QTY      PCT     37.5
----------H2O---------               CF0070 PUMP P         PSID    9.3
CF0009 WASTE         PCT      24.4   SF0262 RAD IN T       °F      76.5
       WASTE         LB       13.9   SF0263 RAD OUT T      °F      46.2
CF0010 POTABLE       PCT      104.1  CF0073 STEAM P        PSIA    .2460
       POTABLE       LB       37.5   CF0071 EVAP OUT T     °F      66.3
CF0460 URINE NOZ T     °F     72     CF0120 H2O RES        PSIA    25.8
CF0461 H2O NOZ T       °F     72     TOTAL  FC CUR         AMPS    71.58
-------CRYO SUPPLY---------O2-1----O2-2-------H2-1------H2-2---
SC0037-38-39-40    P PSIA     879      996      224.2      233.6
SC0032-33-30-32   QTY  PCT    76.83    47.04    73.24      74.03
SC0041-42-43-44    T   °F    -190     -329     -417       -416
                  QTY  LBS    248.5    260.0    20.61      20.83
```

Figure 10.3 The moment of the explosion

Computer based systems do not have to be designed clumsily, obscuring the perception of events in the underlying system being monitored. Poor feedback about events and change is a design failure. Research on representation design has developed techniques to create effective representations – views that make us smart (Zhang, 1997; Flach, Hancock, Caird, and Vicente, 1996).

HOW COMPUTER-BASED ARTIFACTS SHAPE COGNITION AND COLLABORATION 165

Four seconds after the explosion

```
LM1839                          CSM ECS CRYOTAB                      0613
CTE 055:54:56 (55.915)   GET 055:54:58 (55.916)            SITE GDS09
------LIFE SUPPORT-----              ----PRIMARY COOLANT----
GF3571 LM CABIN P   PSIA             CF0019 ACCUM QTY      PCT    34.4
CF0001 CABIN P      PSIA    5.1      CF0016 PUMP P         PSID   46.7
CF0012 SUIT P       PSIA    4.1      SF0260 RAD IN T       °F     73.8
CF0003 SUIT ΔP      IN H2O -1.64
CF0015 COMP ΔP      P PSID  0.30     CF0020 RAD OUT T      °F       35
CF0006 SURGE P      P PSIA  891      CF1081 EVAPINT        °F     45.9
       SURGE QTY    LB      3.67     CF0017 STEAM T        °F     64.4
   O2  TK1 CAP ΔP   PSID    -109     CF0034 STEAM P        PSIA    .161
   O2  TK2 CAP ΔP   PSID    -872     CF0018 EVAP OUT T     °F     44.2

CF0036 O2 MAN P     PSIA    105      SF0266 RAD VLV1/2               ONE
CF0035 O2 FLOW      LB/HR   0.181    CF0175 GLY FLO        LB/HR 214.6
CF0008 SUIT T       °F      50.8
CF0002 CABIN T      °F      65       ---SECONDARY COOLANT---
CF0005 CO2 PP       MMHG    1.5      CF0072 ACCUM QTY      PCT    36.8
----------H2O----------              CF0070 PUMP P         PSID    9.3
CF0009 WASTE        PCT     24.8     SF0262 RAD IN T       °F     76.5
       WASTE        LB      13.9     SF0263 RAD OUT T      °F     46.2
CF0010 POTABLE      PCT     104.1    CF0073 STEAM P        PSIA   .2460
       POTABLE      LB      37.5     CF0071 EVAP OUT T     °F     66.3
CF0460 URINE NOZ T  °F      72       CF0120 H2O RES        PSIA   25.8
CF0461 H2O NOZ T    °F      72       TOTAL  FC CUR         AMPS  81.45
-------CRYO SUPPLY----------02-1----02-2--------H2-1------H2-2---
SC0037-38-39-40     P   PSIA   782      19       224.2         233.6
SC0032-33-30-32 QTY     PCT   78.04    47.04     73.64         74.03
SC0041-42-43-44 T       °F    -190     84        -417          -416
                QTY     LBS  252.4    260.0      20.72         20.83
```

Figure 10.4 Four seconds after the explosion. Reconstruction of the same computer display shown in Figure 10.3 during the pressure transient caused by the explosion. Note oxygen tank 2 pressure at this exact point is 996 psi

Four minutes after the explosion

```
LM1839                          CSM ECS CRYOTAB                      0613
CTE 055:58:24 (55.930)   GET 055:58:26 (55.974)            SITE GDS09
------LIFE SUPPORT-----              ----PRIMARY COOLANT----
GF3571 LM CABIN P   PSIA             CF0019 ACCUM QTY      PCT    35.6
CF0001 CABIN P      PSIA    5.1      CF0016 PUMP P         PSID   46.9
CF0012 SUIT P       PSIA    4.1      SF0260 RAD IN T       °F     56.4
CF0003 SUIT ΔP      IN H2O -1.60
CF0015 COMP ΔP      P PSID  0.30     CF0020 RAD OUT T      °F       34
CF0006 SURGE P      P PSIA  891      CF1081 EVAPINT        °F     44.6
       SURGE QTY    LB      3.67     CF0017 STEAM T        °F     64.4
   O2  TK1 CAP ΔP   PSID    -514     CF0034 STEAM P        PSIA    .161
   O2  TK2 CAP ΔP   PSID    -872     CF0018 EVAP OUT T     °F     44.0

CF0036 O2 MAN P     PSIA    105      SF0266 RAD VLV1/2               ONE
CF0035 O2 FLOW      LB/HR   0.181    CF0175 GLY FLO        LB/HR 211.9
CF0008 SUIT T       °F      50.5
CF0002 CABIN T      °F      65       ---SECONDARY COOLANT---
CF0005 CO2 PP       MMHG    1.5      CF0072 ACCUM QTY      PCT    36.8
----------H2O----------              CF0070 PUMP P         PSID    9.3
CF0009 WASTE        PCT     25.6     SF0262 RAD IN T       °F     76.8
       WASTE        LB      14.3     SF0263 RAD OUT T      °F     47.4
CF0010 POTABLE      PCT     104.1    CF0073 STEAM P        PSIA   .2460
       POTABLE      LB      37.5     CF0071 EVAP OUT T     °F     65.7
CF0460 URINE NOZ T  °F      73       CF0120 H2O RES        PSIA   25.8
CF0461 H2O NOZ T    °F      76       TOTAL  FC CUR         AMPS  61.29
-------CRYO SUPPLY----------02-1----02-2--------H2-1------H2-2---
SC0037-38-39-40     P   PSIA   377      19       228.7         236.6
SC0032-33-30-32 QTY     PCT   74.81   -103       74.05         -1.24
SC0041-42-43-44 T       °F    -195    -329       -417          -427
                QTY     LBS  241.9    260.0      20.84         -0.35
```

Figure 10.5 Four minutes after the explosion: Four seconds after the explosion pressure had dropped to 19 psi, where it remained four minutes later

To briefly illustrate some of these techniques, consider the oxygen tank explosion in *Apollo 13*. Recall the controlle's nightmare. It is reported the recurring dream ended only when a new version of the dream came – he looked at the critical digitals "before the bang and saw the pressure rising. ... Then the tank blew, and he saw the pressure drop and told Flight exactly what had happened" (Murray and Cox, 1989, p. 407).

Data are informative based on relationships to other data, relationships to larger frames of reference, and relationships to the interests and expectations of the observer. In this case what frames of reference would capture the meaningful relationships that allow the observer to "see" the event of the explosion and its consequences from the sensor data? Time is one possibility, as Figure 10.6 illustrates, that allows representation of events as temporally extended behaviors of a device or process.

Depicting relationships in a frame of reference

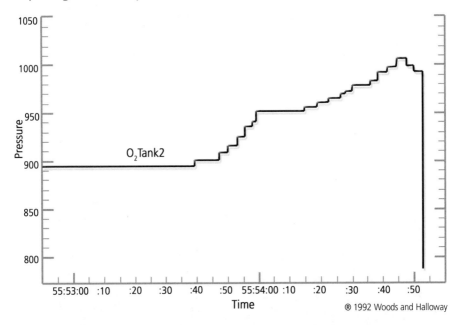

Figure 10.6 Depicting relationships in a frame of reference. One such frame of reference is a space defined by the value against time. This frame of reference expresses a set of relationships about temporal events in the cryogenics systems and their associated functions

Another challenge is the context sensitivity problem – what is interesting depends on the context in which it occurs. The current value (906, or 986, or 19) is just a piece of data. This datum is informative based on its relationship to other values and the expectations of the observer. What related data give meaning to values of oxygen tank pressure? In the case of a thermodynamic parameter like pressure these related data are targets, limits, and set points. What should be the value in this operating context? What are relevant limits

Putting data into context

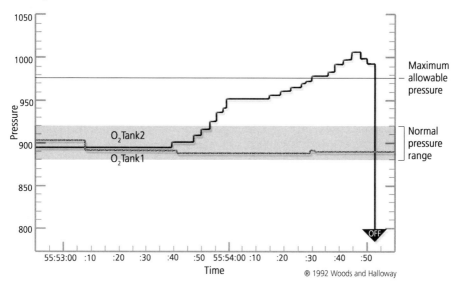

Figure 10.7 Putting data into context: Examples of how pressure data is informative based on its relationship to other values and the expectations of the observer

or regions which should be avoided? What are critical values that trigger or deactivate automated systems?

Effective representations highlight "operationally interesting" changes or sequences of behavior, for example, capturing approach to a limit, recovery from a disturbance, a deteriorating situation, or what event will happen next if things continue moving in the same direction. The sequence of events in this part of the incident are:

○ stable, normal
○ drifting high
○ abnormally high and increasing,
○ and most critically
○ rapid depressurization: off scale high followed immediately by off scale low
○ depressurized (off-scale low).

The temporal frame of reference captures these different events in the relationship between current value and landmarks and past values. The event of rapidly changing from abnormally high to off-scale low is now directly visible (see Figure 10.8).

Since meaning lies in contrasts – some departure from a reference or expected course, the representation should highlight expected paths. In the original display, expectations of how the data should behave are contained entirely in the head of the controller. The data only becomes informative when it is known that:

Highlighting events and contrasts

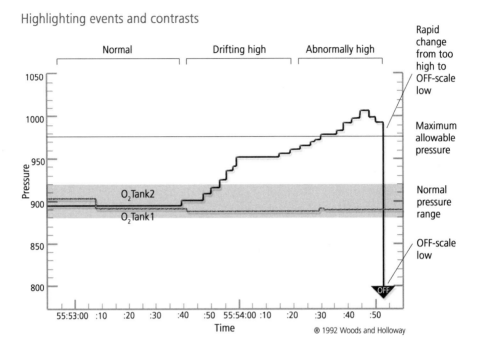

Figure 10.8 Highlighting events and contrasts: The event of rapidly changing from abnormally high to off-scale low in the behavior of oxygen tank pressure is now directly visible in contrast to expected behavior during the critical seconds of the *Apollo 13* mission

O   Both oxygen tanks should be reading approximately the same value.
O   The pressure in this part of the flight should be approximately between 880 and 920 psi.
O   A value of 19 psi is off-scale low, meaning that the tank is depressurized, therefore holding no oxygen, and no longer supporting its functions.

Representing change and events is critical because the computer medium affords the possibility of dynamic reference – the behavior of the representation can refer to the structure and behavior of the referent objects and processes. As the flight controller monitored his systems he only saw the end result – a depressurized system based on the displayed value (19), his domain knowledge, his knowledge of the context, and his expectations. In contrast, notice how the frame of reference annotated with landmarks allows the story of the explosion to become a visible event, rather than a conclusion derived from a deliberative chain of reasoning. One drawback of this representation is its demand on screen real estate. It is important to understand the tradeoffs involved in choosing to annotate data values vs. placing every value in context. At one extreme, annotating raw data values requires the user to keep the data history and expectations in the head. At the other extreme, displaying every value in a graphical format over time requires the user to navigate between many displays. Balancing these tradeoffs is a difficult design challenge that is part of thinking about a display system (Woods and Watts, 1997).

Given that the computer representation is free from the physical constraints acting on the referent objects, support for event perception in the computer medium requires the designer to actively identify operationally interesting changes or sequences of behavior and to actively develop representations that highlight these events to the observer. The default representations and displays typically available to practitioners do not make interesting events directly visible. Rather, practitioners are forced into a serial deliberative mode of cognition to abstract change and events from the displayed data.

ESCAPING FROM ATTRIBUTIONS OF HUMAN ERROR VERSUS OVER-AUTOMATION

This section has only briefly introduced a few of the typical ways that technology is used clumsily. It is very easy to mis-interpret deficiencies such as these. The misinterpretation is to think these deficiencies mean that incidents are "caused" by "over-automation" as opposed to the more typical reaction after-the-fact that claims incidents are "caused" by "human error." We remain locked into a mindset of thinking that technology and people are independent components – either this electronic box failed or that human box failed. This has been the case in aviation as stakeholders have struggled with data, incidents and accidents that involve clumsy automation. For example,

- human error – "clear misuse of automation … contributed to crashes of trouble free aircraft" (La Burthe, 1997) or to
- over-automation – "statements made by … Human Factors specialists against automation 'per se' " (La Burthe, 1997).

This opposition is a profound misunderstanding of the factors that influence human performance. One commentator on human-computer interaction makes this point by defining the term interface as "an arbitrary line of demarcation set up in order to apportion the blame for malfunctions" (Kelly-Bootle, 1995, p. 101).

The careful analysis of incidents involving intelligent and automated computer systems, the Second Story, reveals a breakdown in coordination between people and technology. It is extremely important to highlight this: clumsy use of technology is about miscoordination between the human and machine portion of a single ensemble (Christoffersen and Woods, 2002; Klein et al., 2004). Again, technological artifacts can enhance human expertise or degrade it, "make us smart" or "make us dumb." Automation and people have to coordinate as a joint system, a single team. Breakdowns in this team's coordination is an important path towards disaster. The potential for constructive progress comes from developing better ways to coordinate the human and machine team – human-centered design (Norman, 1990b; Hutchins, 1995a, 1995b; Billings 1996; Winograd and Woods, 1997; Woods and Hollnagel, 2006).

The above results do not imply that we must abandon advanced computer-based technology. Technology is just a kind of power. These cases illustrate the difference between using the power of technology clumsily and skillfully from the point of view of the operational system. This point will become clearer in next chapter as we take one human-computer problem – mode error, and explore it in detail as a model of the impact flow diagram.

11

MODE ERROR IN SUPERVISORY CONTROL

This chapter provides a comprehensive overview of mode error and possible countermeasures (mode error was introduced as a classic pattern behind the label "human error"). Mode errors occur when an operator executes an intention in a way that would be appropriate if the device were in one configuration (one mode) when it is in fact in another configuration. Note that mode errors are not simply just human error or a machine failure. Mode errors are a kind of human-machine system breakdown in that it takes both a user who loses track of the current system configuration, and a system that interprets user-input differently depending on the current mode of operation. The potential for mode error increases as a consequence of a proliferation of modes and interactions across modes without changes to improve the feedback to users about system state and activities.

Several studies have shown how multiple modes can lead to erroneous actions and assessments, and several design techniques have been proposed to reduce the chances for mode errors (for examples of the former see Lewis and Norman (1986); for examples of the latter see Monk, (1986) and Sellen, Kurtenbach, and Buxton (1992)). These studies also illustrate how evaluation methods can and should be able to identify computerized devices which have a high potential for mode errors. To understand the potential for mode error one needs to analyze the computer-based device in terms of what modes and mode transitions are possible, the context of how modes may come into effect in different situations, and how the mode of the device is represented in these different contexts.

Characteristics of the computer medium (e.g., virtuality) and characteristics of and pressures on design make it easy for developers to proliferate modes and to create more complex interactions across modes. The result is new opportunities for mode errors to occur and new kinds of mode-related problems. This chapter provides an overview of the current knowledge and understanding of mode error.

THE CLASSIC CONCEPT OF MODE ERROR

The concept of mode error was originally developed in the context of relatively simple computerized devices, such as word processors, used for self-paced tasks where the device only reacts to user inputs and commands. Mode errors in these contexts occur when an intention is executed in a way appropriate for one mode when, in fact, the system is in a different mode. In this case, mode errors present themselves phenotypically as errors of commission. The mode error that precipitated the chain of events leading to the Strasbourg accident, in part, may have been of this form (Monnier, 1992; Lenorovitz, 1992). The pilot appears to have entered the correct digits for the planned descent given the syntactical input requirements (33 was entered, intended to mean an angle of descent of 3.3 degrees); however, the automation was in a different descent mode which interpreted the entered digits as a different instruction (as meaning a rate of descent of 3,300 feet per minute). Losing track of which mode the system was in, mode awareness, seems to have been another component of mode error in this case.

In one sense a mode error involves a breakdown in going from intention to specific actions. But in another sense a breakdown in situation assessment has occurred – the practitioner has lost track of device mode. One part of this breakdown in situation assessment seems to be that device or system modes tend to change at a different rhythm relative to other user inputs or actions. Mode errors emphasize that the consequences of an action depend on the context in which it is carried out. On the surface, the operator's intention and the executed action(s) appear to be in correspondence; the problem is that the meaning of action is determined by another variable – the system's mode status.

Designers should examine closely the mode characteristics of computerized devices and systems for the potential for creating this predictable form of human-computer breakdown. Multiple modes shape practitioner cognitive processing in two ways. First, the use of multiple modes increases memory and knowledge demands – one needs to know or remember the effects of inputs and the meanings of indications in the various modes. Second, it increases demands on situation assessment and awareness. The difficulty of these demands is conditional on how the interface signals device mode (observability) and on characteristics of the distributed set of agents who manage incidents. The difficulty of keeping track of which mode the device is in also varies depending on the task context (time-pressure, interleaved multiple tasks, workload).

Design countermeasures to the classic mode problems are straightforward in principle:

- Eliminate unnecessary modes (in effect, recognize that there is a cost in operability associated with adding modes for flexibility, marketing, and other reasons).
- Look for ways to increase the tolerance of the system to mode error. Look at specific places where mode errors could occur and (since these are errors of commission) be sure that (a) there is a recovery window before negative consequences accrue and (b) that the actions are reversible.
- Provide better indications of mode status and better feedback about mode changes.

MODE ERROR AND AUTOMATED SYSTEMS

Human supervisory control of automated resources in event-driven task domains is a quite different type of task environment compared to the applications in the original research on mode error. Automation is often introduced as a resource for the human supervisor, providing him with a large number of modes of operation for carrying out tasks under different circumstances. The human's role is to select the mode best suited to a particular situation.

However, this flexibility tends to create and proliferate modes of operation which create new cognitive demands on practitioners. Practitioners must know more – both about how the system works in each different mode and about how to manage the new set of options in different operational contexts. New attentional demands are created as the practitioner must keep track of which mode the device is in, both to select the correct inputs when communicating with the automation, and to track what the automation is doing now, why it is doing it, and what it will do next. These new cognitive demands can easily congregate at high-tempo and high-criticality periods of device use thereby adding new workload at precisely those time periods where practitioners are most in need of effective support systems.

These cognitive demands can be much more challenging in the context of highly automated resources. First, the flexibility of technology allows automation designers to develop much more complicated systems of device modes. Designers can provide multiple levels of automation and more than one option for many individual functions. As a result, there can be quite complex interactions across the various modes including "indirect" mode transitions. As the number and complexity of modes increase, it can easily lead to separate fragmented indications of mode status. As a result, practitioners have to examine multiple displays, each containing just a portion of the mode status data, to build a complete assessment of the current mode configuration.

Second, the role and capabilities of the machine agent in human-machine systems have changed considerably. With more advanced systems, each mode itself is an automated function which, once activated, is capable of carrying out long sequences of tasks autonomously in the absence of additional commands from human supervisors. For example, advanced cockpit automation can be programmed to automatically control the aircraft shortly after takeoff through landing. This increased capability of the automated resources themselves creates increased delays between user input and feedback about system behavior. As a result, the difficulty of error or failure detection and recovery goes up and inadvertent mode settings and transitions may go undetected for long periods. This allows for mode errors of omission (i.e., failure to intervene) in addition to mode errors of commission in the context of supervisory control.

Third, modes can change in new ways. Classically, mode changes only occurred as a reaction to direct operator input. In advanced technology systems, mode changes can occur indirectly based on situational and system factors as well as operator input. In the case of highly automated cockpits, for example, a mode transition can occur as an immediate consequence of pilot input. But it can also happen when a preprogrammed intermediate target (e.g., a target altitude) is reached or when the system changes its mode to prevent the pilot from putting the aircraft into an unsafe configuration.

This capability for "indirect" mode changes, independent of direct and immediate instructions from the human supervisor, drives the demand for mode awareness. Mode awareness is the ability of a supervisor to track and to anticipate the behavior of automated systems. Maintaining mode awareness is becoming increasingly important in the context of supervisory control of advanced technology which tends to involve an increasing number of interacting modes at various levels of automation to provide the user with a high degree of flexibility. Human supervisors are challenged to maintain awareness of which mode is active and how each active or armed mode is set up to control the system, the contingent interactions between environmental status and mode behavior, and the contingent interactions across modes. Mode awareness is crucial for any users operating a multi-mode system that interprets user input in different ways depending on its current status.

CASE 11.1   MODE ERROR AT SEA (LÜTZHÖFT AND DEKKER, 2002)

A cruise ship named the *Royal Majesty*, sailing from Bermuda to Boston in the Summer of 1995. It had more than 1,000 people onboard. Instead of Boston, the *Royal Majesty* ended up on a sandbank close to the Massachusetts shore. Without the crew noticing, it had drifted 17 miles off course during a day and a half of sailing. Investigators discovered afterward that the ship's autopilot had defaulted to DR (Dead Reckoning) mode (from NAV, or Navigation mode) shortly after departure. DR mode does not compensate for the effects of wind and other drift (waves, currents), which NAV mode does. A northeasterly wind pushed the ship steadily off its course, to the side of its intended track. The U.S. National Transportation Safety Board investigation into the accident judged that "despite repeated indications, the crew failed to recognize numerous opportunities to detect that the vessel had drifted off track" (NTSB, 1997, p. 34).

A crew ended up 17 miles off track, after a day and a half of sailing. How could this happen? As said before, hindsight makes it easy to see where people were, versus where they thought they were. In hindsight, it is easy to point to the cues and indications that these people should have picked up in order to update or correct or even form their understanding of the unfolding situation around them. Hindsight has a way of exposing those elements that people missed, and a way of amplifying or exaggerating their importance. The key question is not why people did not see what we now know was important. The key question is how they made sense of the situation the way they did. What must the crew in question at the time have seen? How could they, on the basis of their experiences, construct a story that was coherent and plausible? What were the processes by which they became sure that they were right about their position, and how did automation help with this?

The *Royal Majesty* departed Bermuda, bound for Boston at 12:00 noon on the 9th of June 1995. The visibility was good, the winds light, and the sea calm. Before departure the navigator checked the navigation and communication equipment. He found it in "perfect operating condition."

About half an hour after departure the harbor pilot disembarked and the course was set toward Boston.

Just before 13:00 there was a cutoff in the signal from the GPS (Global Positioning System) antenna, routed on the fly bridge (the roof of the bridge), to the receiver – leaving the receiver without satellite signals. Post-accident examination showed that the antenna cable had separated from the antenna connection. When it lost satellite reception, the GPS promptly defaulted to dead reckoning (DR) mode. It sounded a brief aural alarm and displayed two codes on its tiny display: DR and SOL. These alarms and codes were not noticed. (DR means that the position is estimated, or deduced, hence "ded," or now "dead," reckoning. SOL means that satellite positions cannot be calculated.) The ship's autopilot would stay in DR mode for the remainder of the journey.

Why was there a DR mode in the GPS in the first place, and why was a default to that mode neither remarkable, nor displayed in a more prominent way on the bridge? When this particular GPS receiver was manufactured (during the 1980s), the GPS satellite system was not as reliable as it is today. The receiver could, when satellite data was unreliable, temporarily use a DR mode in which it estimated positions using an initial position, the gyrocompass for course input and a log for speed input. The GPS thus had two modes, normal and DR. It switched autonomously between the two depending on the accessibility of satellite signals.

By 1995, however, GPS satellite coverage was pretty much complete, and had been working well for years. The crew did not expect anything out of the ordinary. The GPS antenna was moved in February, because parts of the superstructure occasionally would block the incoming signals, which caused temporary and short (a few minutes, according to the captain) periods of DR navigation. This was to a great extent remedied by the antenna move, as the cruise line's electronics technician testified. People on the bridge had come to rely on GPS position data and considered other systems to be backup systems. The only times the GPS positions could not be counted on for accuracy were during these brief, normal episodes of signal blockage. Thus, the whole bridge crew was aware of the DR-mode option and how it worked, but none of them ever imagined or were prepared for a sustained loss of satellite data caused by a cable break – no previous loss of satellite data had ever been so swift, so absolute, and so long-lasting.

When the GPS switched from normal to DR on this journey in June 1995, an aural alarm sounded and a tiny visual mode annunciation appeared on the display. The aural alarm sounded like that of a digital wristwatch and was less than a second long. The time of the mode change was a busy time (shortly after departure), with multiple tasks and distracters competing for the crew's attention. A departure involves complex maneuvering, there are several crew members on the bridge, and there is a great deal of communication. When a pilot disembarks, the operation is time constrained and risky. In such situations, the aural signal could easily have been drowned out. No one was expecting a reversion to DR mode, and thus the visual indications were not seen either. From the insider perspective, there was no alarm, as there was not going to be a mode default. There was neither a history, nor an expectation of its occurrence.

Yet even if the initial alarm was missed, the mode indication was continuously available on the little GPS display. None of the bridge crew saw it, according to their testimonies. If they had seen it, they knew what it meant, literally translated – dead reckoning means no satellite fixes. But as we saw before, there is a crucial difference between data that in hindsight can be shown to have been available and data that were observable at the time. The indications on the little display (DR and SOL) were placed between two rows of numbers (representing the ship's latitude and longitude) and were about one sixth the size of those numbers. There was no difference in the size and character of the position indications after the switch to DR. The size of the display screen was about 7.5 by 9 centimeters, and the receiver was placed at the aft part of the bridge on a chart table, behind a curtain. The location is reasonable, because it places the GPS, which supplies raw position data, next to the chart, which is normally placed on the chart table. Only in combination with a chart do the GPS data make sense, and furthermore the data were forwarded to the integrated bridge system and displayed there (quite a bit more prominently) as well.

For the crew of the *Royal Majesty*, this meant that they would have to leave the forward console, actively look at the display, and expect to see more than large digits representing the latitude and longitude. Even then, if they had seen the two-letter code and translated it into the expected behavior of the ship, it is not a certainty that the immediate conclusion would have been "this ship is not heading towards Boston anymore," because temporary DR reversions in the past had never led to such dramatic departures from the planned route. When the officers did leave the forward console to plot a position on the chart, they looked at the display and saw a position, and nothing but a position, because that is what they were expecting to see. It is not a question of them not attending to the indications. They were attending to the indications, the position indications, because plotting the position it is the professional thing to do. For them, the mode change did not exist.

But if the mode change was so nonobservable on the GPS display, why was it not shown more clearly somewhere else? How could one small failure have such an effect – were there no backup systems? The *Royal Majesty* had a modern integrated bridge system, of which the main component was the navigation and command system (NACOS).

The NACOS consisted of two parts, an autopilot part to keep the ship on course and a map construction part, where simple maps could be created and displayed on a radar screen. When the *Royal Majesty* was being built, the NACOS and the GPS receiver were delivered by different manufacturers, and they, in turn, used different versions of electronic communication standards.

Due to these differing standards and versions, valid position data and invalid DR data sent from the GPS to the NACOS were both labeled with the same code (GP). The installers of the bridge equipment were not told, nor did they expect, that (GP-labeled) position data sent to the NACOS would be anything but valid position data. The designers of the NACOS expected that if invalid data were received, they would have another format. As a result, the GPS used the same data label for valid and invalid data, and thus the autopilot could not distinguish between them. Because the NACOS could not detect that the GPS data was invalid, the ship sailed on an autopilot that was using estimated positions until a few minutes before the grounding.

A principal function of an integrated bridge system is to collect data such as depth, speed, and position from different sensors, which are then shown on a centrally placed display to provide the officer of the watch with an overview of most of the relevant information. The NACOS on the *Royal Majesty* was placed at the forward part of the bridge, next to the radar screen. Current technological systems commonly have multiple levels of automation with multiple mode indications on many displays. An better design strategy is to collect these in the same place and another solution is to integrate data from many components into the same display surface. This presents an integration problem for shipping in particular, where quite often components are delivered by different manufacturers.

The centrality of the forward console in an integrated bridge system also sends the implicit message to the officer of the watch that navigation may have taken place at the chart table in times past, but the work is now performed at the console. The chart should still be used, to be sure, but only as a backup option and at regular intervals (customarily every half-hour or every hour). The forward console is perceived to be a clearing house for all the information needed to safely navigate the ship.

As mentioned, the NACOS consisted of two main parts. The GPS sent position data (via the radar) to the NACOS in order to keep the ship on track (autopilot part) and to position the maps on the radar screen (map part). The autopilot part had a number of modes that could be manually selected: NAV and COURSE. NAV mode kept the ship within a certain distance of a track, and corrected for drift caused by wind, sea, and current. COURSE mode was similar but the drift was calculated in an alternative way. The NACOS also had a DR mode, in which the position was continuously estimated. This backup calculation was performed in order to compare the NACOS DR with the position received from the GPS. To calculate the NACOS DR position, data from the gyrocompass and Doppler log were used, but the initial position was regularly updated with GPS data. When the *Royal Majesty* left Bermuda, the navigation officer chose the NAV mode and the input came from the GPS, normally selected by the crew during the 3 years the vessel had been in service.

If the ship had deviated from her course by more than a preset limit, or if the GPS position had differed from the DR position calculated by the autopilot, the NACOS would have sounded an aural and clearly shown a visual alarm at the forward console (the position-fix alarm). There were no alarms because the two DR positions calculated by the NACOS and the GPS were identical. The NACOS DR, which was the perceived backup, was using GPS data, believed to be valid, to refresh its DR position at regular intervals. This is because the GPS was sending DR data, estimated from log and gyro data, but labeled as valid data. Thus, the radar chart and the autopilot were using the same inaccurate position information and there was no display or warning of the fact that DR positions (from the GPS) were used. Nowhere on the integrated display could the officer on watch confirm what mode the GPS was in, and what effect the mode of the GPS was having on the rest of the automated system, not to mention the ship.

In addition to this, there were no immediate and perceivable effects on the ship because the GPS calculated positions using the log and the gyrocompass. It cannot be expected that a crew should become suspicious of the fact that the ship actually is keeping her speed and course. The combination of a busy departure, an unprecedented event (cable break) together with a nonevent (course keeping), and the change of the locus of navigation (including the intrasystem communication difficulties) shows that it made sense, in the situation and at the time, that the crew did not know that a mode change had occurred.

Even if the crew did not know about a mode change immediately after departure, there was still a long voyage at sea ahead. Why did none of the officers check the GPS position against another source, such as the Loran-C receiver that was placed close to the GPS? (Loran-C is a radio navigation system that relies on land-based transmitters.) Until the very last minutes before the grounding, the ship did not act strangely and gave no reason for suspecting that anything was amiss. It was a routine trip, the weather was good and the watches and watch changes uneventful.

Several of the officers actually did check the displays of both Loran and GPS receivers, but only used the GPS data (because those had been more reliable in their experience) to plot positions on the paper chart. It was virtually impossible to actually observe the implications of a difference between Loran and GPS numbers alone. Moreover, there were other kinds of cross-checking. Every hour, the position on the radar map was checked against the position on the paper chart, and cues in the world (e.g., sighting of the first buoy) were matched with GPS data. Another subtle reassurance to officers must have been that the master on a number of occasions spent several minutes checking the position and progress of the ship, and did not make any corrections.

Before the GPS antenna was moved, the short spells of signal degradation that led to DR mode also caused the radar map to jump around on the radar screen (the crew called it "chopping") because the position would change erratically. The reason chopping was not observed on this particular occasion was that the position did not change erratically, but in a manner consistent with dead reckoning. It is entirely possible that the satellite signal was lost before the autopilot was switched on, thus causing no shift in position. The crew had developed a strategy to deal with this occurrence in the past. When the position-fix alarm sounded, they first changed modes (from NAV to COURSE) on the autopilot and then they acknowledged the alarm. This had the effect of stabilizing the map on the radar screen so that it could be used until the GPS signal returned. It was an unreliable strategy, because the map was being used without knowing the extent of error in its positioning on the screen. It also led to the belief that, as mentioned earlier, the only time the GPS data were unreliable was during chopping. Chopping was more or less alleviated by moving the antenna, which means that by eliminating one problem a new pathway for accidents was created. The strategy of using the position-fix alarm as a safeguard no longer covered all or most of the instances of GPS unreliability.

This locally efficient procedure would almost certainly not be found in any manuals, but gained legitimacy through successful repetition becoming common practice over time. It may have sponsored the belief that a stable map is a good map, with the crew concentrating on the visible signs instead of being wary of the errors hidden below the surface. The chopping problem had been resolved for about four months, and trust in the automation had grown.

Especially toward the end of the journey, there appears to be a larger number of cues that retrospective observers would see as potentially revelatory of the true nature of the situation. The first officer could not positively identify the first buoy that marked the entrance of the Boston sea lanes (Such lanes form a separation scheme delineated on the chart to keep meeting and crossing traffic at a safe distance and to keep ships away from dangerous areas). A position error was still not suspected, even with the vessel close to the shore. The lookouts reported red lights and later blue and white water, but the second officer did not take any action. Smaller ships in the area broadcast warnings on the radio, but nobody on the bridge of the *Royal Majesty* interpreted those to concern their vessel.

The second officer failed to see the second buoy along the sea lanes on the radar, but told the master that it had been sighted.

The first buoy ("BA") in the Boston traffic lanes was passed at 19:20 on the 10th of June, or so the chief officer thought (the buoy identified by the first officer as the BA later turned out to be the "AR" buoy located about 15 miles to the west-southwest of the BA). To the chief officer, there was a buoy on the radar, and it was where he expected it to be, it was where it should be. It made sense to the first officer to identify it as the correct buoy because the echo on the radar screen coincided with the mark on the radar map that signified the BA. Radar map and radar world matched. We now know that the overlap between radar map and radar return was a mere stochastic fit. The map showed the BA buoy, and the radar showed a buoy return. A fascinating coincidence was the sun glare on the ocean surface that made it impossible to visually identify the BA. But independent cross-checking had already occurred: The first officer probably verified his position by two independent means, the radar map and the buoy.

The officer, however, was not alone in managing the situation, or in making sense of it. An interesting aspect of automated navigation systems in real workplaces is that several people typically use it, in partial overlap and consecutively, like the watch-keeping officers on a ship. At 20:00 the second officer took over the watch from the chief officer. The chief officer must have provided the vessel's assumed position, as is good watch-keeping practice. The second officer had no reason to doubt that this was a correct position. The chief officer had been at sea for 21 years, spending 30 of the last 36 months onboard the *Royal Majesty*. Shortly after the takeover, the second officer reduced the radar scale from 12 to 6 nautical miles. This is normal practice when vessels come closer to shore or other restricted waters. By reducing the scale, there is less clutter from the shore, and an increased likelihood of seeing anomalies and dangers.

When the lookouts later reported lights, the second officer had no expectation that there was anything wrong. To him, the vessel was safely in the traffic lane. Moreover, lookouts are liable to report everything indiscriminately; it is always up to the officer of the watch to decide whether to take action. There is also a cultural and hierarchical gradient between the officer and the lookouts; they come from different nationalities and backgrounds. At this time, the master also visited the bridge and, just after he left, there was a radio call. This escalation of work may well have distracted the second officer from considering the lookouts' report, even if he had wanted to.

After the accident investigation was concluded, it was discovered that two Portuguese fishing vessels had been trying to call the *Royal Majesty* on the radio to warn her of the imminent danger. The calls were made not long before the grounding, at which time the *Royal Majesty* was already 16.5 nautical miles from where the crew knew her to be. At 20:42, one of the fishing vessels called, "fishing vessel, fishing vessel call cruise boat," on channel 16 (an international distress channel for emergencies only). Immediately following this first call in English the two fishing vessels started talking to each other in Portuguese. One of the fishing vessels tried to call again a little later, giving the position of the ship he was calling. Calling on the radio without positively identifying the intended receiver can lead to mix-ups. In this case, if the second officer heard the first English call and the ensuing conversation, he most likely disregarded it since it seemed to be two other vessels talking to each other. Such an interpretation makes sense: If one ship calls without identifying the intended receiver, and another ship responds and consequently engages the first caller in conversation, the communication loop is closed. Also, as the officer was using the 6-mile scale, he could not see the fishing vessels on his radar. If he had heard the second call and checked the position, he might well have decided that the call was not for him, as it appeared that he was far from that position. Whomever the fishing ships were calling, it could not have been him, because he was not there.

At about this time, the second buoy should have been seen and around 21:20 it should have been passed, but was not. The second officer assumed that the radar map was correct when it showed that they were on course. To him the buoy signified a position, a distance traveled in the traffic lane, and reporting that it had been passed may have amounted to the same thing as reporting that they had passed the position it was (supposed to have been) in. The second officer did not, at this time, experience an accumulation of anomalies, warning him that something was going wrong. In his view, this buoy, which was perhaps missing or not picked up by the radar, was the first anomaly, but not perceived as a significant one. The typical Bridge Procedures Guide says that a master should be called when (a) something unexpected happens, (b) when something expected does not happen (e.g., a buoy), and (c) at any other time of uncertainty. This is easier to write than it is to apply in practice, particularly in a case where crew members do not see what they expected to see. The NTSB report, in typical counterfactual style, lists at least five actions that the officer should have taken. He did not take any of these actions, because he was not missing opportunities to avoid the grounding. He was navigating the vessel normally to Boston.

The master visited the bridge just before the radio call, telephoned the bridge about one hour after it, and made a second visit around 22:00. The times at which he chose to visit the bridge were calm and uneventful, and did not prompt the second officer to voice any concerns, nor did they trigger the master's interest in more closely examining the apparently safe handling of the ship. Five minutes before the grounding, a lookout reported blue and white water. For the second officer, these indications alone were no reason for taking action. They were no warnings of anything

> about to go amiss, because nothing was going to go amiss. The crew knew where they were. Nothing in their situation suggested to them that they were not doing enough or that they should question the accuracy of their awareness of the situation.
>
> At 22:20 the ship started to veer, which brought the captain to the bridge. The second officer, still certain that they were in the traffic lane, believed that there was something wrong with the steering. This interpretation would be consistent with his experiences of cues and indications during the trip so far. The master, however, came to the bridge and saw the situation differently, but was too late to correct the situation. The *Royal Majesty* ran aground east of Nantucket at 22:25, at which time she was 17 nautical miles from her planned and presumed course. None of the over 1,000 passengers were injured, but repairs and lost revenues cost the company $7 million.

The complexity of modes, interactions across modes, and indirect mode changes create new paths for errors and failures. No longer are modes only selected and activated through deliberate explicit actions. Rather, modes can change as a side effect of other practitioner actions or inputs depending on the system status at the time. The active mode that results may be inappropriate for the context, but detection and recovery can be very difficult in part due to long time-constant feedback loops.

An example of such an inadvertent mode activation contributed to a major accident in the aviation domain (the Bangalore accident) (Lenorovitz, 1990). In that case, one member of the flight crew put the automation into a mode called OPEN DESCENT during an approach without realizing it. In this mode aircraft speed was being controlled by pitch rather than thrust (controlling by thrust was the desirable mode for this phase of flight, that is, in the SPEED mode). As a consequence, the aircraft could not sustain both the glide path and maintain the pilot-selected target speed at the same time. As a result, the flight director bars commanded the pilot to fly the aircraft well below the required descent profile to try to maintain airspeed. It was not until 10 seconds before impact that the other crew member discovered what had happened, too late for them to recover with engines at idle. How could this happen?

One contributing factor in this accident may have been several different ways of activating the OPEN DESCENT mode (i.e., at least five). The first two options involve the explicit manual selection of the OPEN DESCENT mode. In one of these two cases, the activation of this mode is dependent upon the automation being in a particular state.

The other three methods of activating the OPEN DESCENT mode are indirect in the sense of not requiring any explicit manual mode selection. They are related to the selection of a new target altitude in a specific context or to protections that prevent the aircraft from exceeding a safe airspeed. In this case, for example, the fact that the automation was in the ALTITUDE ACQUISITION mode resulted in the activation of OPEN DESCENT mode when the pilot selected a lower altitude. The pilot may not have been aware of the fact that the aircraft was within 200 feet of the previously entered target altitude (which is the definition of ALTITUDE ACQUISITION mode). Consequently, he may not have expected that the selection of a lower altitude at that point would result in a mode transition. Because he did not expect any mode change, he may not have closely monitored his mode annunciations, and hence missed the transition.

Display of data can play an important role when user-entered values are interpreted differently in different modes. In the following example, it is easy to see how this may result in unintended system behavior. In a current highly automated or "glass cockpit" aircraft, pilots enter a desired vertical speed or a desired flight path angle via the same display. The interpretation of the entered value depends on the active display mode. Although the different targets differ considerably (for example, a vertical speed of 2,500 feet vs. a flight path angle of 2.5 degrees), these two targets on the display look almost the same (see Figure 11.1). The pilot has to know to pay close attention to the labels that indicate mode status. He has to remember the indications associated with different modes, when to check for the currently active setting, and how to interpret the displayed indications. In this case, the problem is further aggravated by the fact that feedback about the consequences of an inappropriate mode transition is limited. The result is a cognitively demanding task; the displays do not support a mentally economical, immediate apprehension of the active mode.

Figure 11.1 Example of multiple modes and the potential for mode error on the flight deck of an advanced technology aircraft. The same entry means different things in different modes. Finding the difference in mode indications is left as an exercise to the reader

Coordination across multiple team members is another important factor contributing to mode error in advanced systems. Tracking system status and behavior becomes more difficult if it is possible for other users to interact with the system without the need for consent by all operators involved (the indirect mode changes are one human-machine example of this).

This problem is most obvious when two experienced operators have developed different strategies of system use. When they have to cooperate, it is particularly difficult for them to maintain awareness of the history of interaction with the system which may determine the effect of the next system input. In addition, the design of the interface to the automation may suppress important kinds of cues about the activities of other team members.

The demands for mode awareness are critically dependent on the nature of the interface between the human and machine agents (and as pointed out above between human agents as well). If the computerized device also exhibits another of the HCI problems we noted earlier – not providing users with effective feedback about changes in the state of a device, automated system, or monitored process – then losing track of which mode the device is in may be surprisingly easy, at least in higher workload periods.

Some of these factors contributed to a fatal test flight accident that involved one of the most advanced automated aircraft in operation at the time (*Aviation Week and Space Technology*, April 3, April 10, and April 17, 1995). The test involved checking how the automation could handle a simulated engine failure at low altitude under extreme flight conditions, and it was one of a series of tests being performed one after the other. The flight crew's task in this test was to set up the automation to fly the aircraft and to stop one engine to simulate an engine failure.

There were a number of contributing factors identified in the accident investigation report. During takeoff, the co-pilot rotated the aircraft rather rapidly which resulted in a slightly higher than planned pitch angle (a little more than 25 degrees) immediately after takeoff. At that point, the autopilot was engaged as planned for this test. Immediately following the autopilot engagement, the captain brought the left engine to idle power and cut off one hydraulic system to simulate an engine failure situation. The automation flew the aircraft into a stall. The flight crew recognized the situation and took appropriate recovery actions, but there was insufficient time (altitude) to recover from the stall before a crash killing everyone aboard the aircraft.

How did this happen? When the autopilot was selected, it immediately engaged in an altitude capture mode because of the high rate of climb and because the pilot selected a rather low level-off altitude of 2,000 ft. As a consequence, the automation continued to try to follow an altitude acquisition path even when it became impossible to achieve it (after the captain had brought the left engine to idle power).

The automation performs protection functions which are intended to prevent or recover from unsafe flight attitudes and configurations. One of these protection functions guards against excessive pitch which results in too low an airspeed and a stall. This protection is provided in all automation configurations except one – the very altitude acquisition mode in which the autopilot was operating.

At the same time, because the pitch angle exceeded 25 degrees at that point, the declutter mode of the Primary Flight Display activated. This means that all indications of the active mode configuration of the automation (including the indication of the altitude capture mode) were hidden from the crew because they had been removed from the display for simplification.

Ultimately, the automation flew the aircraft into a stall, and the crew was not able to recover in time because the incident occurred at low altitude.

Clearly, a combination of factors contributed to this accident and was cited in the report of the accident investigation. Included in the list of factors are the extreme conditions under which the test was planned to be executed, the lack of pitch protection in the altitude acquisition mode, and the inability of the crew to determine that the automation had entered that particular mode or to assess the consequences. The time available for the captain to react (12 seconds) to the abnormal situation was also cited as a factor in this accident.

A more generic contributing factor in this accident was the behavior of the automation which was highly complex and difficult to understand. These characteristics made it hard for the crew to anticipate the outcome of the maneuver. In addition, the observability of the system was practically non-existent when the declutter mode of the Primary Flight Display activated upon reaching a pitch angle of more than 25 degrees up.

The above examples illustrate how a variety of factors can contribute to a lack of mode awareness on the part of practitioners. Gaps or misconceptions in practitioners' mental models may prevent them from predicting and tracking indirect mode transitions or from understanding the interactions between different modes. The lack of salient feedback on mode status and transitions (low observability) can also make it difficult to maintain awareness of the current and future system configuration. In addition to allocating attention to the different displays of system status and behavior, practitioners have to monitor environmental states and events, remember past instructions to the system, and consider possible inputs to the system by other practitioners. If they manage to monitor, integrate, and interpret all this information, system behavior will appear deterministic and transparent. However, depending on circumstances, missing just one of the above factors can be sufficient to result in an automation surprise and the impression of an animate system that acts independently of operator input and intention.

As illustrated in the above sections, mode error is a form of human-machine system breakdown. As systems of modes become more interconnected and more autonomous, new types of mode-related problems are likely, unless the extent of communication between man and machine changes to keep pace with the new cognitive demands.

THE "GOING SOUR" SCENARIO

Incidents that express the consequences of clumsy use of technology such as mode error follow a unique signature. Minor disturbances, misactions, miscommunications and miscoordinations seem to be managed into hazard despite multiple opportunities to detect that the system is heading towards negative consequences.

For example, this has occurred in several aviation accidents involving highly automated aircraft (Billings, 1996; Sarter, Woods and Billings, 1997). Against a background of the activities, the flight crew misinstructs the automation (e.g., a mode error). The automation accepts the instructions providing limited feedback confirming only the entries themselves. The flight crew believes they have instructed the automation to do one thing when in fact it will carry out a different instruction. The automation proceeds to fly the aircraft according to these instructions even though this takes the aircraft towards hazard (e.g., off course towards a mountain or descending too rapidly short of the runway). The flight crew, busy with other activities, does not see that the automation is flying the aircraft differently than they had expected and does not see that the aircraft is heading towards hazard until very late. At that point it is too late for the crew to take any effective action to avoid the crash.

When things go wrong in this way, we look back and see a process that gradually went "sour" through a series of small problems. For this reason we call them "going sour" scenarios. This term was originally used to describe scenarios in anesthesiology by Cook, Woods and McDonald (1991). A more elaborate treatment for flightcrew-automation breakdowns in aviation accidents can be found in Sarter, Woods and Billings (1997). In the "going sour" class of accidents, an event occurs or a set of circumstances come together that appear to be minor and unproblematic, at least when viewed in isolation or from hindsight. This event triggers an evolving situation that is, in principle, possible to recover from. But through a series of commissions and omissions, misassessments and miscommunications, the human team or the human-machine team manages the situation into a serious and risky incident or even accident. In effect, the situation is managed into hazard.

After-the-fact, going sour incidents look mysterious and dreadful to outsiders who have complete knowledge of the actual state of affairs. Since the system is managed into hazard, in hindsight, it is easy to see opportunities to break the progression towards disaster. The benefits of hindsight allow reviewers to comment:

o "How could they have missed X, it was the critical piece of information?"
o "How could they have misunderstood Y, it is so logical to us?"
o "Why didn't they understand that X would lead to Y, given the inputs, past instructions and internal logic of the system?"

In fact, one test for whether an incident is a going sour scenario is to ask whether reviewers, with the advantage of hindsight, make comments such as, "All of the necessary data was available, why was no one able to put it all together to see what it meant?" Unfortunately, this question has been asked by a great many of the accident investigation reports of complex system failures regardless of work domain.

Luckily, going sour accidents are relatively rare even in complex systems. The going sour progression is usually blocked because of two factors:

o the problems that can erode human expertise and trigger this kind of scenario are significant only when a collection of factors or exceptional circumstances come together;
o the expertise embodied in operational systems and personnel allows practitioners to avoid or stop the incident progression usually.

COUNTERMEASURES

Mode error illustrates some of the basic strategies researchers have identified to increase the human contribution to safety:

o increase the system's tolerance to errors,

- avoid excess operational complexity,
- evaluate changes in technology and training in terms of their potential to create specific genotypes or patterns of failure,
- increase skill at error detection by improving the observability of state, activities and intentions,
- make intelligent and automated machines team-players,
- invest in human expertise.

We will discuss a few of these in the context of mode error and awareness.

AVOID EXCESS OPERATIONAL COMPLEXITY

Designers frequently do not appreciate the cognitive and operational costs of more and more complex modes. Often, there are pressures and other constraints on designers that encourage mode proliferation. However, the apparent benefits of increased functionality may be more than counterbalanced by the costs of learning about all the available functions, the costs of learning how to coordinate these capabilities in context, and the costs of mode errors. Users frequently cope with the complexity of the modes by "re-designing" the system through patterns of use; for example, few users may actually use more than a small subset of the resident options or capabilities.

Avoiding excess operational complexity is a difficult issue because no single developer or organization decides to make systems complex. But in the pursuit of local improvements or in trying to accommodate multiple customers, systems gradually get more and more complex as additional features, modes, and options accumulate. The cost center for this increase in creeping complexity is the user who must try to manage all of these features, modes and options across a diversity of operational circumstances. Failures to manage this complexity are categorized as "human error." But the source of the problem is not inside the person. The source is the accumulated complexity from an operational point of view. Trying to eliminate "erratic" behavior through remedial training will not change the basic vulnerabilities created by the complexity. Neither will banishing people associated with failures. Instead human error is a symptom of systemic factors. The solutions are system fixes that require change at the blunt end of the system. This coordinated system approach must start with meaningful information about the factors that predictably affect human performance.

Mode simplification in aviation illustrates both the need for change and the difficulties involved. Not all modes are used by all pilots or carriers due to variations in operations and preferences. Still they are all available and can contribute to complexity for operators. Not all modes are taught in transition training; only a set of "basic" modes is taught, and different carriers define different modes as "basic." It is very difficult to get agreement on which modes represent excess complexity and which are essential for safe and efficient operation.

ERROR DETECTION THROUGH IMPROVED FEEDBACK

Research has shown that systems are effective through effective detection and recovery of developing trouble before negative consequences occur. Error detection is improved by providing better feedback, especially feedback about the future behavior of the underlying system or automated systems. In general, increasing complexity can be balanced with improved feedback. Improving feedback is a critical investment area for improving human performance.

One area of need is improved feedback about the current and future behavior of the automated systems. As technological change increases machines' autonomy, authority and complexity, there is a concomitant need to increase observability through new forms of feedback emphasizing an integrated dynamic picture of the current situation, agent activities, and how these may evolve in the future. Increasing autonomy and authority of machine agents without an increase in observability leads to automation surprises. As discussed earlier, data on automation surprises has shown that crews generally do not detect their miscommunications with the automation from displays about the automated system's state, but rather only when system behavior becomes sufficiently abnormal.

This result is symptomatic of low observability where observability is the technical term that refers to the cognitive work needed to extract meaning from available data. This term captures the fundamental relationship among data, observer and context of observation that is fundamental to effective feedback. Observability is distinct from data availability, which refers to the mere presence of data in some form in some location. For human perception, "it is not sufficient to have something in front of your eyes to see it" (O'Regan, 1992, p. 475).

Observability refers to processes involved in extracting useful information (Rasmussen, 1985 first introduced the term referring to control theory). It results from the interplay between a human user knowing when to look for what information at what point in time and a system that structures data to support attentional guidance (Woods, 1995a). The critical test of observability is when the display suite helps practitioners notice more than what they were specifically looking for or expecting (Sarter and Woods, 1997).

One example of displays with very low observability on the current generation of flight decks is the flight-mode annunciations on the primary flight display. These crude indications of automation activities contribute to reported problems with tracking mode transitions. As one pilot commented, "changes can always sneak in unless you stare at it." Simple injunctions for pilots to look closely at or call out changes in these indications generally are not effective ways to redirect attention in a changing environment.

For new display concepts to enhance observability they will need to be:

○ transition-oriented – provide better feedback about events and transitions;
○ future-oriented – the current approach generally captures only the current configuration; the goal is to highlight operationally significant sequences and reveal what will or should happen next and when;
○ pattern-based – practitioners should be able to scan at a glance and quickly pick up possible unexpected or abnormal conditions rather than have to read and integrate each individual piece of data to make an overall assessment.

MECHANISMS TO MANAGE AUTOMATED RESOURCES

Giving users visibility into the machine agent's reasoning processes is only one side of the coin in making machine agents into team players. Without also giving the users the ability to direct the machine agent as a resource in their reasoning processes, the users are not in a significantly improved position. They might be able to say what's wrong with the machine's solution, but remain powerless to influence it in any way other than through manual takeover. The computational power of machine agents provides a great potential advantage, that is, to free users from much of the mundane legwork involved in working through large problems, thus allowing them to focus on more critical high-level decisions. However, in order to make use of this potential, the users need to be given the authority and capabilities to make those decisions. This means giving them control over the problem solution process.

A commonly proposed remedy for this is, in situations where users determine that the machine agent is not solving a problem adequately, to allow users to interrupt the automated agent and take over the problem in its entirety. Thus, the human is cast into the role of critiquing the machine, and the joint system operates in essentially two modes – fully automatic or fully manual. The system is a joint system only in the sense that either a human agent or a machine agent can be asked to deal with the problem, not in the more productive sense of the human and machine agents cooperating in the process of solving the problem. This method, which is like having the automated agent say "either you do it or I'll do it," has many obvious drawbacks. Either the machine does all the job without any benefits of practitioners' information and knowledge, and despite the brittleness of the machine agents, or the user takes over in the middle of a deteriorating or challenging situation without the support of cognitive tools. Previous work in several domains (space operations, electronic troubleshooting, aviation) and with different types of machine agents (expert systems, cockpit automation, flight-path-planning algorithms) has shown that this is a poor cooperative architecture. Instead, users need to be able to continue to work with the automated agents in a cooperative manner by taking control of the automated agents.

Using the machine agent as a resource may mean various things. As for the case of observability, one of the main challenges is to determine what levels and modes of interaction will be meaningful to users. In some cases the users may want to take very detailed control of some portion of a problem, specifying exactly what decisions are made and in what sequence, while in others the users may want only to make very general, high level corrections to the course of the solution in progress. Accommodating all of these possibilities is difficult and requires very careful iterative analysis of the interactions between user goals, situational factors, and the nature of the machine agent.

ENHANCING HUMAN EXPERTISE

Mode awareness also indicates how technology change interacts with human expertise. It is ironic that many industries seem to be reducing the investment in human expertise, at the very time when they claim that human performance is a dominant contributor to accidents.

One of the myths about the impact of new technology on human performance is that as investment in automation increases less investment is needed in human expertise. In fact, many sources have shown how increased automation creates new and different knowledge and skill requirements.

Investigations of mode issue in aviation showed how the complexity of the automated flight deck creates the need for new knowledge about the functions and interactions of different automated subsystems and modes. Data showed how the complexity of the automated flight deck makes it easy for pilots to develop oversimplified or erroneous mental models of the tangled web of automated modes and transition logics. Training departments struggle within very limited time and resource windows to teach crews how to manage the automated systems as a resource in differing flight situations. Many sources have identified incidents when pilots were having trouble getting a particular mode or level of automation to work successfully. In these cases they persisted too long trying to get a particular mode of automation to carry out their intentions instead of switching to another means or a more direct means to accomplish their flight path management goals. For example, after an incident someone may ask those involved, "Why didn't you turn it off?" Response: "It didn't do what it was supposed to, so I tried to get it to do what I had programmed it to do." The new knowledge and skill demands seem to be most relevant in relatively infrequent situations where different kinds of factors push events beyond the routine.

For training managers and departments the result is a great deal of training demands that must be fitted into a small and shrinking training footprint. The new roles, knowledge and skills for practitioners combine with economic pressures to create a training double bind. Trainers may cope with this double bind in many ways. They may focus limited training resources on a basic subset of features, modes and capabilities leaving the remainder to be learned on the job. Another tactic is to teach recipes. But what if the deferred material is the most complicated or difficult to learn? What if people have misconceptions about those aspects of device function – how will they be corrected? What happens when circumstances force practitioners away from the limited subset of device capabilities with which they are most familiar or most practiced? All of these become problems or conditions that can (and have) contributed to failure.

New developments which promise improved training may not be enough by themselves to cope with the training double bind. Economic pressure means that the benefits of improvements will be taken in productivity (reaching the same goal faster) rather than in quality (more effective training). Trying to squeeze more yield from a shrinking investment in human expertise will not help prevent the kinds of incidents and accidents that we label human error after the fact.

12
HOW PRACTITIONERS ADAPT TO CLUMSY TECHNOLOGY

In developing new information technology and automation, the conventional view seems to be that new technology makes for better ways of doing the same task activities. We often act as if domain practitioners were passive recipients of the "operator aids" that the technologist provides for them. However, this view overlooks the fact that the introduction of new technology represents a change from one way of doing things to another.

> The design of new technology is always an intervention into an ongoing world of activity. It alters what is already going on – the everyday practices and concerns of a community of people – and leads to a resettling into new practices. (Flores et al., 1988, p. 154)

For example, Cordesman and Wagner summarize lessons from the use of advanced technology in the Gulf War:

> Much of the equipment deployed in US, other Western, and Saudi forces was designed to ease the burden on the operator, reduce fatigue, and simplify the tasks involved in combat. Instead, these advances were used to demand more from the operator. Almost without exception, technology did not meet the goal of unencumbering the military personnel operating the equipment, due to the burden placed on them by combat. As a result … systems often required exceptional human expertise, commitment, and endurance. … leaders will exploit every new advance to the limit. As a result, virtually every advance in ergonomics was exploited to ask military personnel to do more, do it faster and do it in more complex ways. … One very real lesson of the Gulf War is that new tactics and technology simply result in altering the pattern of human stress to achieve a new intensity and tempo of combat. (Cordesman and Wagner, 1996, p. 25)

Practitioners are not passive in this process of accommodation to change. Rather, they are an active adaptive element in the person-machine ensemble, usually the critical adaptive portion. Multiple studies have shown that practitioners adapt information technology provided for them to the immediate tasks at hand in a locally pragmatic way, usually in ways not anticipated by the designers of the information technology (Roth et al., 1987; Flores et al., 1988; Hutchins, 1990; Cook and Woods, 1996b; Obradovich and Woods, 1996). Tools are shaped by their users. Or, to state the point more completely, artifacts are shaped into tools through skilled use in a field of activity. This process, in which an artifact is shaped by its use, is a fundamental characteristic of the relationship between design and use.

> There is always … a substantial gap between the design or concept of a machine, a building, an organizational plan or whatever, and their operation in practice, and people are usually well able to effect this translation. Without these routine informal capacities most organizations would cease to function. (Hughes, Randall, and Shapiro, 1991)

TAILORING TASKS AND SYSTEMS

Studies have revealed several types of practitioner adaptation to the impact of new information technology. In system tailoring, practitioners adapt the device and context of activity to preserve existing strategies used to carry out tasks (e.g., adaptation focuses on the set-up of the device, device configuration, how the device is situated in the larger context). In task tailoring, practitioners adapt their strategies, especially cognitive and collaborative strategies, for carrying out tasks to accommodate constraints imposed by the new technology.

When practitioners tailor systems, they adapt the device itself to fit their strategies and the demands of the field of activity. For example, in one study (summarized in Chapter 7), practitioners set up the new device in a particular way to minimize their need to interact with the new technology during high criticality and high-tempo periods. This occurred despite the fact that the practitioners' configurations neutralized many of the putative advantages of the new system (e.g., the flexibility to perform greater numbers and kinds of data manipulation). Note that system tailoring frequently results in only a small portion of the "in principle" device functionality actually being used operationally. We often observe operators throw away or alter functionality in order to achieve simplicity and ease of use.

Task tailoring types of adaptations tend to focus on how practitioners adjust their activities and strategies given constraints imposed by characteristics of the device. For example, information systems which force operators to access related data serially through a narrow keyhole instead of in parallel result in new displays and window management tasks (e.g., calling up and searching across displays for related data, decluttering displays as windows accumulate, and so on). Practitioners may tailor the device itself, for example, by trying to configure windows so that related data is available in parallel. However, they may still need to tailor their activities. For example, they may need to learn when to schedule the new decluttering task (e.g., by devising external reminders) to avoid being caught in a high criticality situation where they must reconfigure the display before they can "see" what is going on in the monitored process. A great many user adaptations can be found in observing how people cope with large numbers of displays hidden behind a narrow keyhole. See Woods and Watts (1997) for a review.

PATTERNS IN USER TAILORING

Task and system tailoring represent coping strategies for dealing with clumsy aspects of new technology. A variety of coping strategies employed by practitioners to tailor the system or their tasks have observed (Roth et al., 1987; Cook et al., 1991; Moll van Charante et al., 1993; Sarter and Woods, 1993; Sarter and Woods, 1995; Cook and Woods, 1996b; Obradovich and Woods, 1996). One class of coping behaviors relates to workload management to prevent bottlenecks from occurring at high-tempo periods. For example, we have observed practitioners force device interaction to occur in low-workload periods to minimize the need for interaction at high-workload or high-criticality periods. We have observed practitioners abandon cooperative strategies and switch to single-agent strategies when the demands for communication with the machine agent are high, as often occurs during high criticality and high tempo operations.

Another class of coping strategies relates to spatial organization. We consistently observe users constrain "soft," serial forms of interaction and display into a spatially dedicated default organization.

Another consistent observation is that, rather than exploit device flexibility, we see practitioners externally constrain devices via ad hoc standards. Individuals and groups develop and stick with stereotypical routes or methods to avoid getting lost in large networks of displays, complex menu structures, or complex sets of alternative methods. For example, Figure 12.1 shows about 50 percent of the menu space for a computerized patient-monitoring information system used in operating rooms. We sampled physician interaction with the system for the first three months of its use during cardiac surgery. The highlighted sections of the menu space indicate the options that were actually used by physicians during this time period. This kind of data is typical – to cope with complexity, users throw away functionality to achieve simplicity of use tailored to their perceptions of their needs.

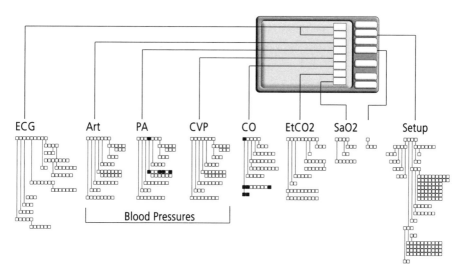

Figure 12.1 How practitioners cope with complexity in computerized devices. This figure illustrates a portion of the menu space for a computerized patient-monitoring information system. The highlighted areas are the items actually used by practitioners during observations of device use in cardiac surgery over three months. Note that the space of possibilities is very large compared with the portion practitioners actually use. (From Cook and Woods, 1994)

Studies of practitioner adaptation to clumsy technology consistently observe users invent "escapes" – ways to abandon high-complexity modes of operation and to retreat to simpler modes of operation when workload gets too high.

Finally, observations indicate that practitioners sometimes learn ways to "trick" automation, for example to silence nuisance alarms. Practitioners appear to do this in an attempt to exercise control over the technology (rather than let the technology control them) and to get the technology to function as a resource or tool for their ends.

The patterns of adaptation noted above represent examples of practitioners' adaptive coping strategies for dealing with clumsy aspects of new technology, usually in response to criteria such as workload, cognitive effort, robustness. Note these forms of tailoring are as much a group as an individual dynamic. Understanding how practitioners adaptively respond to the introduction of new technology and understanding what are the limits of their adaptations are critical for understanding how new automation creates the potential for new forms of error and system breakdown.

BRITTLE TAILORING

Practitioners (commercial pilots, anesthesiologists, nuclear power operators, operators in space control centers, and so on) are responsible, not just for device operation but also for the larger system and performance goals of the overall system. Practitioners tailor their activities to insulate the larger system from device deficiencies and peculiarities of the technology. This occurs, in part, because practitioners inevitably are held accountable for failure to correctly operate equipment, diagnose faults, or respond to anomalies even if the device setup, operation, and performance are ill-suited to the demands of the environment.

However, there are limits to a practitioner's range of adaptability, and there are costs associated with practitioners' coping strategies, especially in non-routine situations when a variety of complicating factors occur. Tailoring can be clever, or it can be brittle. In extreme cases, user adaptations to cope with an everyday glitch can bypass or erode defenses against failure. Reason, working backwards from actual disasters, has noted that such poor adaptations are often a latent condition contributing to disaster (e.g., the capsizing of the ferry *The Herald of Free Enterprise*). Here, we are focusing on the gaps and conflicts in the work environment that lead practitioners to tailor strategies and tasks to be effective in their everyday practice. Adaptations to cope with one glitch can create vulnerabilities with respect to other work demands or situations. Therefore, to be successful in a global sense, adaptation at local levels must be guided by the provision of appropriate criteria, informational and material resources, and feedback. One function of people in more supervisory roles is to coordinate adaptation to recognize and avoid brittle tailoring and to propagate clever ones.

These costs or limits of adaptation represent a kind of an ever-present problem whose effects are visible only when other events and circumstances show up to produce critical incidents. At one point in time, practitioners adapt in effective ways based on the prevailing conditions. However, later, these conditions change in a way that makes the practitioners tailoring ineffective, brittle, or maladaptive.

Clumsy new systems introduce new burdens or complexities for practitioners who adapt their tasks and strategies to cope with the difficulties while still achieving their multiple goals. But incidents can occur that challenge the limits of those adaptations.

Ironically, when incidents occur where those adaptations are brittle, break down, or are maladaptive, investigation stops with the label "human error" (the First Story). When investigations stop at this First Story, the human skills required to cope with the effects of the complexities of the technology remain unappreciated except by other beleaguered practitioners. Paradoxically, practitioners' normally adaptive, coping responses help to hide the corrosive effects of clumsy technology from designers and reviewers. Note the paradox: because practitioners are responsible, they work to smoothly accommodate new technology. As a result, practitioners' work to tailor the technology can make it appear smooth, hiding the clumsiness from designers.

ADAPTATION AND ERROR

There is a fundamental sense in which the adaptive processes which lead to highly skilled, highly robust operator performance are precisely the same as those which lead to failure. Adaptation is basically a process of exploring the space of possible behaviors in search of stable and efficient modes of performance given the demands of the field of practice.

"Error" in this context represents information about the limits of successful adaptation. Adaptation thus relies on the feedback or learning about the system which can be derived from failures, near misses and incidents. Change and the potential for surprise (variability in the world) drive the need for individuals, groups and organizations to adapt (Ashby's law of requisite variety). If the result is a positive outcome, we tend to call it cleverness, skill or foresight and the people or organizations reap rewards. If the result is a negative outcome, we tend to call it "human error" and begin remedial action.

Attempts to eradicate "error" often take the form of policing strict adherence to policies and procedures. This tactic tries to eliminate the consequences of poor adaptations by attempting to drive out all adaptations. Given the inherent variability of real fields of practice, systems must have equivalent adaptive resources. Pressures from the organization to stick to standard procedures when external circumstances require adaptation to achieve goals creates a double bind for practitioners: fail to adapt and goals will not be met but adaptation if unsuccessful will result in sanctions.

To find the Second Story after incidents, we need to understand more about how practitioners adapt tools to their needs and to the constraints of their field of activity. These adaptations may be inadequate or successful, misguided or inventive, brittle or robust, but they are locally rational responses of practitioners to match resources to demands in the pursuit of multiple goals.

HOW DESIGNERS CAN ADAPT

Computer technology offers enormous opportunities. And enormous pitfalls. In the final analysis, the enemy of safety is complexity. In medicine, nuclear power and aviation, among other safety-critical fields, we have learned at great cost that often it is the underlying complexity of operations, and the technology with which to conduct them, that contributes to the human performance problems. Simplifying the operation of the system can do wonders to improve its reliability, by making it possible for the humans in the system to operate effectively and more easily detect breakdowns. Often, we have found that proposals to improve systems founder when they increase the complexity of practice (e.g., Xiao et al., 1996). Adding new complexity to already complex systems rarely helps and can often make things worse.

The search for operational simplicity, however, has a severe catch. The very nature of improvements and efficiency (e.g., in health care delivery) includes, creates, or exacerbates many forms of complexity. Ultimately, success and progress occur through monitoring, managing, taming, and coping with the changing forms of complexity, and not by mandating simple "one size fits all" policies. This has proven true particularly with respect to efforts to introduce new forms and levels of computerization. Improper computerization can simply exacerbate or create new forms of complexity to plague operations.

ADOPT METHODS FOR USE-CENTERED DESIGN OF INFORMATION TECHNOLOGY

Calls for more use of integrated computerized information systems to reduce error could introduce new and predictable forms of error unless there is a significant investment in use-centered design. The concepts and methods for use-centered design are available and are being used everyday in software houses (Carroll and Rosson, 1992; Flach and Dominguez, 1995). Focus groups, cognitive walkthroughs, and interviews are conducted to generate a cognitive task analysis which details the nature of work to be supported by a product (e.g., Garner and Mann, 2003). Iterative usability testing of a system prior to use with a handful of representative users has become a standard, not an exceptional part, of most product development practices. "Out of the box" testing is conducted to elicit feedback on how to improve the initial installation and use of a fielded product.

Building partnerships, creating demonstration projects, and disseminating the techniques for organizations is a significant and rewarding investment to ensure we receive the benefits of computer technology while avoiding designs that induce new errors (Kling, 1996). But there is much more to human-computer interaction than adopting basic techniques like usability testing (Karsh, 2004). Much of the work in human factors concerns how to use the potential of computers to enhance expertise and performance. The key to skillful as opposed to clumsy use of technological possibilities lies in understanding both the factors that lead to expert performance and the factors that challenge expert performance (Feltovich, Ford, and Hoffman, 1997). Once we understand the factors that contribute to expertise and to breakdown, we then will understand how to use the powers of the computer to enhance expertise. This is an example of a more general rule – to understand failure and success, begin by understanding what makes some problems difficult. We can achieve substantial gains by understanding the factors that lead to expert performance and the factors that challenge expert performance. This provides the basis to change the system, for example, through new computer support systems and other ways to enhance expertise in practice.

PART V
REACTIONS TO FAILURE

In the book so far, we have covered many different aspects of research on human error and the evolution of system failures. The results indicate that the story of "human error" is markedly complex because it is always possible to identify multiple contributors to an incident or disaster, each necessary but only jointly sufficient. Furthermore, the story of error is complex because:

○ the human performance in question involves a distributed system of interacting people at the sharp end and organizational elements at the blunt end;
○ the same factors govern the expression of both expertise and error;
○ the context in which incidents evolve plays a major role in human performance at the sharp end;
○ people at the blunt end create dilemmas and shape tradeoffs among competing goals for those at the sharp end; and
○ the way technology is deployed shapes human performance, creating the potential for new forms of error and failure, as well as new forms of expert performance.

In this last part, we explore other important factors that contribute to the complexity of error. A critical factor is the hindsight bias, which demonstrates how the attribution of error after-the-fact is a process of social and psychological judgment rather than an objective conclusion. Instead of using outcome as a criterion for judging performance, we explore the possibility of using standards for evaluating good process. This too, however, requires judgments about the likelihood of particular processes leading to successful outcomes given different features and goals of the field of activity. What dimensions of performance should guide the evaluation, for example, efficiency or robustness; safety or throughput? The loose coupling between process and outcome leaves us with a continuing nagging problem. Defining human error as a form of process defect implies that there exists some criterion or standard against which the activities of the agents in the system have been measured and deemed inadequate. However, what standard should be used to mark a process as deficient? Depending on the standard a reviewer adopts, very different views of error result.

This is a most necessary part of the book. Incidents and accidents always challenge stakeholders' belief in the safety of the system and the adequacy of the defenses and control measures that are in place. After-the-fact stakeholders look back and make judgments about what led to the accident or incident. Lay people, scientists, engineers, managers, and regulators all judge what "caused" the event in question, reactions to failure that are influenced by many factors. As discussed in many places in this book, one of the most critical is that, after an accident, we know the outcome. Knowledge of outcome biases people's judgment about the processes that led up to that outcome. We react, after the fact, as if knowledge of outcome was available to operators as well, and wonder why they didn't see it coming. This oversimplifies and trivializes the situation confronting the practitioners, and masks the processes affecting practitioner behavior before the fact. These processes of social and psychological attribution are such an important obstacle to getting to the second story of systematic factors which predictably shape human performance that the last part of the book is dedicated to them entirely.

One of the themes reverberating throughout this book is that human error represents a symptom rather than a cause. In this view error is, in part, a form of information about the functioning of the system in which those people are embedded. And if errors are a form of information, then they represent a basis for learning about the system, including its human-machine interaction and socio-technical systems. But human and institutional reactions to failure may not always want to see error only as learning opportunities. There are many cases in which stakeholders voice demands for "accountability." An entire chapter of this part is devoted to handling the aftermath of failure, particularly how stakeholders think about balancing learning from errors with accountability for errors. We discuss the problem of the criminalization of error, and outline possible efforts for building a "just culture."

We have a responsibility, driven by the consequences that can accompany failure, to maximize the information value of such potentially expensive feedback. Some researchers believe that one important measure of the reliability of an organization, in the sense of resilience or robustness, is how it responds to evidence of failures. Lower-reliability organizations tend to react with a search for culprits. Achieving greater safety in socio-technical and human-machine systems demands that we look hard, directly, and honestly in every way at incidents, disasters and their precursors. If we label events as "human error" and stop, what have we learned? As the many examples and concepts in this book illustrate, the answer is: very little. The label "human error" is a judgment made in hindsight. Failures occur in systems that people develop and operate for human purposes. Such systems are not and cannot be purely technological; they always involve people at various levels and in various ways. We cannot pretend that technology alone, or more procedures, or better monitoring, or more rules, divorced from the people who develop, shape, and use them in practice, will be enough. Failure and success are both forms of information about the system in which people are embedded. The potential for constructive change lies behind the label "human error," and this final part of the book concludes with a summary of ideas on how to do just that.

13
HINDSIGHT BIAS

ATTRIBUTING SYSTEM FAILURES TO PRACTITIONERS

System failures, near failures, and critical incidents are the usual triggers for investigations of human performance. When critical incidents do occur, human error is often seen as a cause of the poor outcome. In fact, large complex systems can be readily identified by the percentage of critical incidents that are considered to have been "caused" by "human error;" the rate for these systems is typically over 70 percent. The repeated finding of about three-quarters of incidents arising from "human error" has built confidence in the notion that there is a human error problem in these domains. Indeed, the belief that fallible humans are responsible for large system failures has led many system designers to use more and more technology to try to eliminate the human operator from the system or to reduce the operator's possible actions so as to forestall these incidents.

Attributing system failure to the human operators nearest temporally and spatially to the outcome ultimately depends on the judgment by someone that the processes in which the operator engaged were faulty and that these faulty processes led to the bad outcome. Deciding which of the many factors surrounding an incident are important and what level or grain of analysis to apply to those factors is the product of social and psychological processes of causal attribution. What we identify as the cause of an incident depends on what we ourselves have learned previously, where we look, whom we communicate with, on assumed contrast cases or causal background for that exchange, and on the purposes of the inquiry.

For at least four reasons it is not surprising that human operators are blamed for bad outcomes. First, operators are available to blame. Large and intrinsically dangerous systems have a few, well-identified humans at the sharp end. Those humans are closely identified with the system function so that it is unlikely that a bad outcome will occur without having them present. Moreover, these individuals are charged, often formally and institutionally, with ensuring the safe operation as well as the efficient functioning of the system. For any large system failure there will be a human in close temporal and physical relationship to the outcome (e.g., a ship's captain, pilot, air traffic controller, physician, nurse).

The second reason that "human error" is often the verdict after accidents is that it is so difficult to trace backwards through the causal chain of multiple contributors that are involved in system failure (Rasmussen, 1986). It is particularly difficult to construct a sequence that "passes through" humans in the chain, as opposed to stopping at the sharp-end human(s). To construct such a sequence requires the ability to reconstruct, in detail, the cognitive processing of practitioners during the events that preceded the bad outcome. The environment of the large system makes these sorts of reconstructions extremely difficult. Indeed, a major area of research is development of tools to help investigators trace the cognitive processing of operators as they deal with normal situations, with situations at the edges of normality, and with system faults and failures. The incidents described in Part III are unusual in that substantial detail about what happened, what the participants saw and did, was available to researchers. In general, most traces of causality will begin with the outcome and work backwards in time until they encounter a human whose actions seem to be, in hindsight, inappropriate or sub-optimal. Because so little is known about how human operators actually deal with the multiple conflicting demands of large, complex systems, incident analyses rarely demonstrate the ways in which the actions of the operator made sense at the time.

The third reason that "human error" is often the verdict is paradoxical: "human error" is attributed to be the cause of large system accidents because human performance in these complex systems is so good. Failures of these systems are, by almost any measure, rare and unusual events. Most of the system operations go smoothly; incidents that occur do not usually lead to bad outcomes. These systems have come to be regarded as "safe" by design rather than by control. Those closely studying human operations in these complex systems are usually impressed by the fact that the opportunity for large-scale system failures is present all the time and that expert human performance is able to prevent these failures. As the performance of human operators improves and failure rates fall, there is a tendency to regard system performance as a marked improvement in some underlying quality of the system itself, rather than the honing of skills and expertise within the distributed operational system to fine edge. The studies of aircraft carrier flight operations by Rochlin et al., (1987) point out that the qualities of human operators are crucial to maintaining system performance goals and that, by most measures, failures should be occurring much more often than they do. As consumers of the products from large complex systems such as health care, transportation, and defense, society is lulled by success into the belief that these systems are intrinsically low-risk and that the expected failure rate should be zero. Only catastrophic failures receive public attention and scrutiny. The remainder of the system operation is generally regarded as unflawed because of the low overt failure rate, even though there are many incidents that could become overt failures. Thorough accident analyses often indicate that there were precursor events or "dress rehearsals" that preceded an accident.

This ability to trace backwards with the advantage of hindsight is the fourth major reason that human error is so often the verdict after accidents. Studies have consistently shown that people have a tendency to judge the quality of a process by its outcome. Information about outcome biases their evaluation of the process that was followed.

Also, people have a tendency to "consistently exaggerate what could have been anticipated in foresight" (Fischhoff, 1975). Typically, hindsight bias in evaluations makes it seem that participants failed to account for information or conditions that "should have been obvious"(when someone claims that something "should have been obvious" hindsight bias is virtually always present) or behaved in ways that were inconsistent with the (now known to be) significant information. Thus, knowledge of a poor outcome biases the reviewer towards attributing failures to system operators. But to decide what would be "obvious" to practitioners in the unfolding problem requires investigating many factors about the evolving incident, the operational system and its organizational context such as the background of normal occurrences, routine practices, knowledge factors, attentional demands, strategic dilemmas, and other factors.

The psychological and social processes involved in judging whether or not a human error occurred is critically dependent on knowledge of the outcome, something that is impossible before the fact. Indeed, it is clear from the studies of large system failures that hindsight bias is the greatest obstacle to evaluating the performance of humans in complex systems.

THE BIASING EFFECT OF OUTCOME KNOWLEDGE

Outcome knowledge influences our assessments and judgments of past events. These hindsight or outcome biases have strong implications for how we study and evaluate accidents, incidents, and human performance.

Whenever one discusses "human error," one should distinguish between outcome failures and defects in the problem-solving process. Outcome failures are defined in terms of a categorical shift in consequences on some performance dimension. Generally, these consequences are directly observable. Outcome failures necessarily are defined in terms of the language of the domain, for example for anesthesiology sequelae such as neurological deficit, reintubation, myocardial infarction within 48 hours, or unplanned ICU admission. Military aviation examples of outcome failures include an unfulfilled mission goal, a failure to prevent or mitigate the consequences of some system failure on the aircraft, or a failure to survive the mission. An outcome failure provides the impetus for an accident investigation.

Process defects, on the other hand, are departures from some standard about how problems should be solved. Generally, the process defect, if uncorrected, would lead to, or increase the risk of, some type of outcome failure. Process defects can be defined in domain terms. For example in anesthesiology, some process defects may include insufficient intravenous access, insufficient monitoring, regional versus general anesthetic, and decisions about canceling a case. They may also be defined psychologically in terms of deficiencies in some cognitive function: for example activation of knowledge in context, mode errors, situation awareness, diagnostic search, and goal tradeoffs.

People have a tendency to judge a process by its outcome. In the typical study, two groups are asked to evaluate human performance in cases with the same descriptive facts but with the outcomes randomly assigned to be either bad or neutral. Those with knowledge of a poor outcome judge the same decision or action more severely. This is referred to as the outcome bias (Baron and Hershey, 1988) and has been demonstrated with practitioners in different domains. For example, Caplan, Posner, and Cheney (1991) found an inverse relationship between the severity of outcome and anesthesiologists' judgments of the appropriateness of care. The judges consistently rated the care in cases with bad outcomes as substandard while viewing the same behaviors with neutral outcomes as being up to standard even though the care (that is, the preceding human acts) were identical. Similarly, Lipshitz (1989) found the outcome bias when middle rank officers evaluated the decisions made by a hypothetical officer. Lipshitz (1989) points out that "judgment by outcomes is a fact of life for decision makers in politics and organizations." In other words, the label "error" tends to be associated with negative outcomes.

It may seem reasonable to assume that a bad outcome stemmed from a bad decision, but information about the outcome is actually irrelevant to the judgment of the quality of the process that led to that outcome (Baron and Hershey, 1988). The people in the problem do not intend to produce a bad outcome (Rasmussen et al., 1987). Practitioners at the sharp end are responsible for action when the outcome is in doubt and consequences associated with poor outcomes are highly negative. If they, like their evaluators, possessed the knowledge that their process would lead to a bad outcome, then they would use this information to modify how they handled the problem. Ultimately, the distinction between the evaluation of a decision process and evaluation of an outcome is important to maintain because good decision processes can lead to bad outcomes and good outcomes may still occur despite poor decisions.

Other research has shown that once people have knowledge of an outcome, they tend to view the outcome as having been more probable than other possible outcomes. Moreover, people tend to be largely unaware of the modifying effect of outcome information on what they believe they could have known in foresight. These two tendencies collectively have been termed the hindsight bias. Fischhoff (1975) originally demonstrated the hindsight bias in a set of experiments that compared foresight and hindsight judgments concerning the likelihood of particular socio-historical events. Basically, the bias has been demonstrated in the following way: participants are told about some event, and some are provided with outcome information. At least two different outcomes are used in order to control for one particular outcome being a priori more likely. Participants are then asked to estimate the probabilities associated with the several possible outcomes. Participants given the outcome information are told to ignore it in coming up with their estimates, that is, "to respond as if they had not known the actual outcome," or in some cases are told to respond as they think others without outcome knowledge would respond. Those participants with the outcome knowledge judge the outcomes they had knowledge about as more likely than the participants without the outcome knowledge.

The hindsight bias has proven to be robust; it has been demonstrated for different types of knowledge: episodes, world facts (e.g., Wood, 1978; Fischhoff, 1977), and in some real-world settings. For example, several researchers have found that medical practitioners exhibited a hindsight bias when rating the likelihood of various diagnoses (cf., Fraser, Smith, and Smith, 1992).

Experiments on the hindsight bias have shown that: (a) people overestimate what they would have known in foresight, (b) they also overestimate what others knew in foresight (Fischhoff, 1975), and (c) they actually misremember what they themselves knew in foresight

(Fischhoff and Beyth, 1975). This misremembering may be linked to the work on reconstructive memory, in which a person's memories can be changed by subsequent information, for example, leading questions may change eyewitnesses' memories (Loftus, 1979).

Fischhoff (1975) postulated that outcome knowledge is immediately assimilated with what is already known about the event. A process of retrospective sense-making may be at work in which the whole event, including outcome, is constructed into a coherent whole. This process could result in information that is consistent with the outcome being given more weight than information inconsistent with it.

> It appears that when we receive outcome knowledge, we immediately make sense out of it by integrating it into what we already know about the subject. Having made this reinterpretation, the reported outcome now seems a more or less inevitable outgrowth of the reinterpreted situation. "Making sense" out of what we are told about the past is, in turn, so natural that we may be unaware that outcome knowledge has had any effect on us. … In trying to reconstruct our foresightful state of mind, we will remain anchored in our hindsightful perspective, leaving the reported outcome too likely looking. (Fischhoff, 1982, p. 343)

It may be that retrospective outsiders (people who observe and judge practitioners' performance from the outside and from hindsight) rewrite the story so that the information is causally connected to the outcome. A study by Wasserman, Lempert, and Hastie (1991) supports this idea. They found that people exhibit more of a hindsight bias when they are given a causal explanation for the outcome than when the outcome provided is due to a chance event (but see Hasher, Attig, and Alba, 1981, for an alternative explanation; see Hawkins and Hastie, 1990, for a summary).

Taken together, the outcome and hindsight biases have strong implications for error analyses.

- Decisions and actions having a negative outcome will be judged more harshly than if the same process had resulted in a neutral or positive outcome. We can expect this result even when judges are warned about the phenomenon and have been advised to guard against it (Fischhoff, 1975, 1982).
- Retrospectively, outsiders will tend to believe that people involved in some incident knew more about their situation than they actually did. Judges will tend to think that people should have seen how their actions would lead up to the outcome failure. Typical questions a person exhibiting the hindsight bias might ask are: "Why didn't they see what was going to happen? It was so obvious!" Or, "How could they have done X? It was clear it would lead to Y!"

Hence it is easy for observers after the fact to miss or underemphasize the role of cognitive, design, and organizational factors in incident evolution. For example, a mode error was probably an important contributor to the Strasbourg crash of an Airbus A-320. As we have seen, this error form is a human-machine system breakdown that is tied to design problems. Yet people rationalize that mode error does not imply the need for design modifications:

> While you can incorporate all the human engineering you want in an aircraft, it's not going to work if the human does not want to read what is presented to him, and verify that he hasn't made an error. (Remarks by Y. Benoist, Director of Flight Safety, Airbus Industry, 1992)

Similarly, in the aftermath of the AT&T's Thomas Street telecommunication outage in 1991, it was easy to focus on individuals at the sharp end and ignore the larger organizational factors.

> It's terrible the incident in New York was (pause) all avoidable. The alarms were disarmed; no one paid attention to the alarms that weren't disarmed; that doesn't have anything to do with technology, that doesn't have anything to do with competition, it has to do with common sense and attention to detail." (Remarks by Richard Liebhaber of MCI commenting on AT&T's Thomas Street outage occurred on September 7, 1991; from MacNeil-Lehrer Report, PBS)

In this case, as in others, hindsight biases the judgment of the commentator. A detailed examination of the events leading up to the Thomas Street outage clearly shows how the alarm issue is, in part, a red herring and clearly implicates failures in the organization and management of the facility (see FCC, 1991).

In effect, judges will tend to simplify the problem-solving situation that was actually faced by the practitioner. The dilemmas facing the practitioner *in situ*, the uncertainties, the tradeoffs, the attentional demands, and the double binds, all may be under-emphasized when an incident is viewed in hindsight. A consideration of practitioners' resources and the contextual and task demands that impinge on them is crucial for understanding the process involved in the incident and for uncovering process defects.

In summary, these biases play a role in how practitioners' actions and decisions are judged after the fact. The biases illustrate that attributing human error or other causes (e.g., software error) for outcomes is a psychological and social process of judgment. These biases can lead us to summarize the complex interplay of multiple contributors with simple labels such as "lack of attention" or "willful disregard." They can make us miss the underlying factors which could be changed to improve the system for the future, for example lack of knowledge or double binds induced by competing goals. Furthermore, the biases illustrate that the situation of an evaluator after-the-fact who does not face uncertainty, risk, and who possesses knowledge of outcome is fundamentally different from that of a practitioner in an evolving problem.

So whenever you hear someone say (or feel yourself tempted to say) something like: "Why didn't they see what was going to happen? It was so obvious!" or "How could they have done *X*? It was clear it would lead to *Y*!" Remember that error is the starting point of an investigation; remember that the error investigator builds a model of how the participants behaved in a locally rational way given the knowledge, attentional demands, and strategic factors at work in that particular field of activity. This is the case regardless of whether one is attributing error to operators, designers, or managers. In other words, it is the responsibility of the error investigator to explore how it could have been hard to see what was going to happen or hard to project the consequences of an action. This does not mean that some assessments or actions are not clearly erroneous. But adoption of the local rationality perspective is important to finding out how and why the erroneous action could have occurred and, therefore, is essential for developing effective countermeasures rather than the usual window dressing of "blame and train," "a little more technology will be enough," or "only follow the rules" recommendations.

Some research has addressed ways to "debias" judges. Simply telling people to ignore outcome information is not effective (Fischhoff, 1975). In addition, telling people about the hindsight bias and to be on guard for it does not seem to be effective (Fischhoff, 1977; Wood, 1978). Strongly discrediting the outcome information can be effective (Hawkins and Hastie, 1990), although this may be impractical for conducting accident analyses.

The method that seems to have had the most success is for judges to consider alternatives to the actual outcome. For example, the hindsight bias may be reduced by asking subjects to explain how each of the possible outcomes might have occurred (Hoch and Lowenstein, 1989). Another relatively successful variant of this method is to ask people to list reasons both for and against each of the possible outcomes (von Winterfeldt and Edwards, 1986; Fraser et al, 1992). This technique is in the vein of a Devil's Advocate approach, which may be one way to guard against a variety of breakdowns in cognitive systems (Schwenk and Cosier, 1980).

This is an example of the general problem solving strategy of considering alternatives to avoid premature closure (Patterson et al., 2001; Zelik et al., 2010).

This work has implications for debiasing judges in accident analysis. But first we need to ask the basic question: What standard of comparison should we use to judge processes (decisions and actions) rather than outcomes?

STANDARDS FOR ASSESSING PROCESSES RATHER THAN OUTCOMES

We have tried to make clear that one of the recurring problems in studying error is a confusion over whether the label is being used to indicate that an outcome failure occurred or that the process used is somehow deficient. The previous section showed that outcome knowledge biases judgments about the processes that led to that outcome. But it seems common sense that some processes are better than others for maximizing the chances of achieving good outcomes regardless of the presence of irreducible uncertainties and risks. And it seems self-evident that some processes are deficient with respect to achieving good outcomes – e.g., relevant evidence may not be considered, meaningful options may not be entertained, contingencies may not have been thought through. But how do we evaluate processes without employing outcome information? How do we know that a contingency should have been thought through except through experience? This is especially difficult given the infinite variety of the real world, and the fact that all systems are resource-constrained. Not all possible evidence, all possible hypotheses, or all possible contingencies can be entertained by limited resource systems. So the question is: what standards can be used to determine when a process is deficient?

There is a loose coupling between process and outcome – not all process defects are associated with bad outcomes, and good process cannot guarantee success given irreducible uncertainties, time pressure, and limited resources. But poor outcomes are relatively easy to spot and to aggregate in terms of the goals of that field of activity (e.g., lives lost, radiation exposure, hull losses, reduced throughput, costs, lost hours due to injuries). Reducing bad outcomes generally is seen as the ultimate criterion for assessing the effectiveness of changes to a complex system. However, measuring the reliability of a complex, highly-coupled system in terms of outcomes has serious limitations. One has to wait for bad outcomes (thus one has to experience the consequences). Bad outcomes may be rare (which is fortunate, but it also means that epidemiological approaches will be inappropriate). It is easy to focus on the unique and local aspects of each bad outcome obscuring larger trends or risks. Bad outcomes involve very many features, factors, and facets: Which were critical? Which should be changed?

If we try to measure the processes that lead to outcomes, we need to define some standard about how to achieve or how to maximize the chances for successful outcomes given the risks, uncertainties, tradeoffs, and resource limitations present in that field of activity. The rate of process defects may be much more frequent than the incidence of overt system failures. This is so because the redundant nature of complex systems protects against many defects. It is also because the systems employ human operators whose function is, in part, to detect such process flaws and adjust for them before they produce bad outcomes.

Process defects can be specified locally in terms of the specific field of activity (e.g., these two switches are confusable). But they also can be abstracted relative to models of error and system breakdown (this erroneous action or system failure is an instance of a larger pattern or syndrome – mode error, latent failures, and so on). This allows one to use individual cases of erroneous actions or system breakdown, not as mere anecdotes or case studies, but rather as individual observations that can be compared, contrasted, and combined to look for, explore, or test larger concepts. It also allows for transfer from one specific setting to another to escape the overwhelming particularity of cases.

STANDARDS FOR EVALUATING GOOD PROCESS

But specifying a process as defective in some way requires an act of judgment about the likelihood of particular processes leading to successful outcomes given different features of the field of activity. What dimensions of performance should guide the evaluation, for example efficiency or robustness; safety or throughput? This loose coupling between process and outcome leaves us with a continuing nagging problem. Defining human error as a form of process defect implies that there exists some criterion or standard against which the activities of the agents in the system have been measured and deemed inadequate. However, what standard should be used to mark a process as deficient? And depending on the standard a reviewer adopts, very different views of error result.

We do not think that there can be a single and simple answer to this question. Given this, we must be very clear about what standards are being used to define "error" in particular studies or incidents; otherwise, we greatly retard our ability to engage in a constructive and empirically grounded debate about error. All claims about when an action or assessment is erroneous in a process sense should be accompanied with an explicit statement of the standard used for defining departures from good process.

One kind of standard about how problems should be handled is a normative model of task performance. This method requires detailed knowledge about precisely how problems should be solved, that is, nearly complete and exhaustive knowledge of the way in which the system works. Such knowledge is, in practice, rare. At best, some few components of the larger system can be characterized in this exhaustive way. As a result, normative models rarely exist for complex fields of activity where bad outcomes have large consequences. There are great questions surrounding how to transfer normative models developed for much simpler situations to these more complex fields of activity (Klein et al., 1993). For example, laboratory-based normative models may ignore the role of time or may assume resource unlimited cognitive processing.

Another standard is the comparison of actual behavior to standard operating procedures or other norms deemed relevant to a profession (e.g., standards of care, policies). These practices are mostly compilations of rules and procedures that are acceptable behaviors for a variety of situations. They include various protocols (e.g., the Advanced Cardiac Life Support protocol for cardiac arrest), policies (e.g., it is the policy of the hospital to have informed consent from all patients prior to beginning an anesthetic), and procedures (e.g., the chief resident calls the attending anesthesiologist to the room before beginning the anesthetic, but after all necessary preparations have been made).

Using standard procedures as a criterion may be of limited value because they are codified in ways that ignore the real nature of the domain. It is not unusual, for example, to have a large body of rules and procedures that are not followed because to do so would make the system intolerably inefficient. The "work to rule" method used by unions to produce an unacceptable slowdown of operations is an example of the way in which reference to standards is unrealistic. In this technique, the workers perform their tasks to an exact standard of the existing rules, and the system performance is so degraded by the extra steps required to conform to all the rules that it becomes non-functional (e.g., see Hirschhorn, 1993).

Standard procedures are severely limited as a criterion because procedures are underspecified and therefore too vague to use for evaluation. For example, one senior anesthesiologist replied, when asked about the policy of the institution regarding the care for emergent caesarean sections, "our policy is to do the right thing." This seemingly curious phrase in fact sums up the problem confronting those at the sharp end of large, complex systems. It recognizes that it is impossible to comprehensively list all possible situations and appropriate responses because the world is too complex and fluid. Thus the person in the situation is required to account for the many factors that are unique to that situation. What sounds like a nonsense phrase is, in fact, an expression of the limitations that apply to all structures of rules, regulations and policies (cf. for example, Suchman, 1987; Roth et al., 1987; Woods and Hollnagel, 2006).

One part of this is that standard procedures underspecify many of the activities and the concomitant knowledge and cognitive factors required to go from a formal statement of a plan to a series of temporally structured activities in the physical world (e.g., Roth et al., 1987; Suchman, 1987). As Suchman puts it, plans are resources for action – an abstraction or representation of physical activity; they cannot, for both theoretical and practical reasons, completely specify all activity.

In general, procedural rules are underspecified and too vague to be used for evaluation if one cannot to determine the adequacy of performance before the fact. Thus, procedural rules such as "the anesthetic shall not begin until the patient has been properly prepared for surgery" or "stop all unnecessary pumps" are underspecified. The practitioner on the scene must use contextual information to define when this patient is "properly prepared" or what pumps are "unnecessary" at this stage of a particular nuclear power-plant incident. Ultimately, it is the role of the human at the sharp end to resolve incompleteness, apparent contradictions, and conflicts in order to satisfy the goals of the system.

A second reason for the gap between formal descriptions of work and the actual work practices is that the formal descriptions underestimate the dilemmas, interactions between constraints, goal conflicts, and tradeoffs present in the actual workplace (e.g., Cook et al., 1991a; Hirschhorn, 1993). In these cases, following the rules may, in fact, require complex judgments as illustrated in the section on double binds. Using standard procedures as a criterion for error may hide the larger dilemma created by organizational factors while providing the administrative hierarchy the opportunity to assign blame to operators after accidents (e.g., see Lauber, 1993 and the report on the aircraft accident at Dryden, Ontario; Moshansky, 1992).

Third, formal descriptions tend to focus on only one agent or one role within the distributed cognitive system. The operator's tasks in a nuclear power plant are described in terms of the assessments and actions prescribed in the written procedures for handling emergencies. But this focuses attention only on how the board operators (those who manipulate the controls) act during "textbook" incidents. Woods has shown through several converging studies of actual and simulated operator decision-making in emergencies that the operational system for handling emergencies involves many decisions, dilemmas, and other cognitive tasks that are not explicitly represented in the procedures (see Woods et al., 1987, for a summary). Emergency operations involve many people in different roles in different facilities beyond the control room. For example, operators confront decisions about whether the formal plans are indeed relevant to the actual situation they are facing, and decisions about bringing additional knowledge sources to bear on a problem.

All these factors are wonderfully illustrated by almost any cognitive analysis of a real incident that goes beyond textbook cases. One of these is captured by a study of one type of incident in nuclear power plants (see Roth et al., 1992). In this case, in hindsight, there is a procedure that identifies the kind of problem and specifies the responses to this particular class of faults. However, handling the incident is actually quite difficult. First, as the situation unfolds in time, the symptoms are similar to another kind of problem with its associated procedures (i.e., the incident has a garden path quality; there is a plausible but erroneous initial assessment). The relationship between what is seen, the practitioners' expectations, and other possible trajectories is critical to understanding the cognitive demands, tasks, and activities in that situation. Second, the timing of events and the dynamic inter-relationships among various processes contain key information for assessing the situation. This temporally contingent data is not well represented within a static plan even if its significance is recognized by the procedure writers. Ultimately, to handle this incident, the operators must step outside the closed world defined by the procedure system.

Standard practices and operating procedures may also miss the fact that for realistically complex problems there is often no one best method. Rather, there is an envelope containing multiple paths, each of which can lead to a satisfactory outcome (Rouse et al., 1984; Woods et al., 1987). Consider the example of an incident scenario used in a simulation study of cognition on the flight deck in commercial aviation (Sarter and Woods, 1993; note that the simulated scenario was based, in part, on an actual incident). To pose a diagnostic problem with certain characteristics (e.g., the need to integrate diverse data, the need to recall and re-interpret past data in light of new developments, and so on), the investigators set up a series of events that would lead to the loss of one engine and two hydraulic systems (a combination that requires the crew to land the aircraft as soon as possible). A fuel tank is underfuelled at the departure airport, but the crew does not realize this, as the fuel gauge for that tank has been declared inoperative by maintenance. For aircraft at that time, there were standards for fuel management, that is, how to feed fuel from the different fuel tanks to the engines. The investigators expected the crews to follow the standard procedures, which in this context would lead to the engine loss, the loss of one of the hydraulic systems, and the associated cognitive demands. And this is indeed what happened except for one crew. This one flight engineer, upon learning that one of his fuel tank gauges would be inoperative throughout the flight, decided to use a non-standard fuel management configuration to ensure that, just in case of any other troubles, he would not lose an engine or risk a hydraulic overheat. In other words, he anticipated some of the potential interactions between the lost indication and other kinds of problems that could arise and then shifted from the standard fuel management practices. Through this non-standard behavior, he prevented all of the later problems that the investigators had set up for the crews in the study.

Did this crew member commit an error? If one's criterion is departure from standard practices, then his behavior was "erroneous." If one focuses on the loss of indication, the pilot's adaptation anticipated troubles that might occur and that might be more difficult to recognize given the missing indication. By this criterion, it is a successful adaptation. But what if the pilot had mishandled the non-standard fuel management approach (a possibility since it would be less practiced, less familiar)? What if he had not thought through all of the side effects of the non-standard approach – did the change make him more vulnerable to other kinds of troubles?

Consider another case, this one an actual aviation incident from 1991 (we condensed the following from the Aviation Safety Reporting System's incident report to reduce aviation jargon and to shorten and simplify the sequence of events):

CASE 13.1 CASCADING AUTOMATED WARNINGS

Climbout was normal, following a night heavy weight departure under poor weather conditions, until approximately 24,000 ft when numerous caution/warning messages began to appear on the cockpit's electronic caution and warning system (CRT-based information displays and alarms about the aircraft's mechanical, electric, and engine systems). The first

> of these warning messages was OVHT ENG 1 NAC, closely followed by BLEED DUCT LEAK L, ENG 1 OIL PRESSURE, FLAPS PRIMARY, FMC L, STARTER CUTOUT 1, and others. Additionally, the #1 engine generator tripped off the line (generating various messages), and the #1 engine amber "REV" indication appeared (indicating a #1 engine reverse). In general, the messages indicated a deteriorating mechanical condition of the aircraft. At approximately 26,000 ft, the captain initiated an emergency descent and turnback to the departing airport.
>
> The crew, supported by two augmented crew pilots (i.e., a total of four pilots), began to perform numerous (over 20) emergency checklists (related to the various warnings messages, the need to dump fuel, the need to follow alternate descent procedures and many others). In fact, the aircraft had experienced a serious pylon/wing fire. Significantly, there was no indication of fire in the cockpit information systems, and the crew did not realize that the aircraft was on fire until informed of this by ATC during the landing roll out. The crew received and had to sort out 54 warning messages on the electronic displays, repeated stick shaker activation, and abnormal speed reference data on the primary flight display. Many of these indications were conflicting, leading the crew to suspect number one engine problems when that engine was actually functioning normally. Superior airmanship and timely use of all available resources enabled this crew to land the aircraft and safely evacuate all passengers and crew from the burning aircraft.

The crew successfully handled the incident – the aircraft landed safely. Therefore, one might say that no errors occurred. On the other hand, the crew did not correctly assess the source of the problems, they did not realize that there was a fire until after touchdown, and they suspected number one engine problems when that engine was actually functioning normally. Should these be counted as erroneous assessments? Recall, though, that the display and warning systems presented "an electronic system nightmare" as the crew had to try to sort out an avalanche of low-level and conflicting indications in a very high-workload and highly critical situation. The incident occurred on a flight with two extra pilots aboard (the nominal crew is two). They had to manage many tasks in order to make an emergency descent in very poor weather and with an aircraft in deteriorating mechanical condition. Note the large number of procedures which had to be coordinated and executed correctly. How did the extra crew contribute to the outcome? Would a standard-sized crew have handled the incident as well? These would be interesting questions to pursue using the neutral-practitioner criteria (see the next section).

The above incidents help to exemplify several points. Assessing good or bad process is extremely complex; there are no simple answers or criteria. Standard practices and procedures provide very limited and very weak criteria for defining errors as bad process. What can one do then? It would be easy to point to other examples of cases where commentators would generally agree that the cognitive process involved was deficient on some score. One implication is to try to develop other methods for studying cognitive processes that provide better insights about why systems fail and how they may be changed to produce higher reliability human-machine systems (Rochlin et al., 1987).

NEUTRAL-PRACTITIONER CRITERIA

The practitioners at the "sharp end" are embedded in an evolving context. They experience the consequences of their actions directly or indirectly. They must act under irreducible uncertainty and the ever-present possibility that in hindsight their responses may turn out wrong. As one critical care physician put it when explaining his field of medicine: "We're the ones who have to do something." It is their job to interpret situations that cannot be completely specified in detail ahead of time. Indeed, it is part of practitioners' tacit job description to negotiate the tradeoffs of the moment.

Blessed with the luxury of hindsight, it is easy to lose the perspective of someone embedded in an evolving situation who experiences the full set of interacting constraints that they must act under. But this is the perspective that we must capture if we are to understand how an incident evolved towards disaster. One technique for understanding the situated practitioner represents a third approach to develop a standard of comparison. One could use an empirical approach, one that asks: "What would other similar practitioners have thought or done in this situation?" De Keyser and Woods (1990) called this kind of empirically based comparison the neutral-practitioner criterion. To develop a neutral-practitioner criterion, one collects data to compare practitioner behavior during the incident in question with the behavior of similar practitioners at various points in the evolving incident and in similar or contrasting cases. In practice, the comparison is usually accomplished by using the judgment of similar practitioners about how they would behave under similar circumstances. Neutral-practitioners make judgments or interpretations about the state of the world, relevant possible future event sequences, and relevant courses of action. The question is whether the path taken by the actual problem-solver is one that is plausible to the neutral-practitioners. One key is to avoid contamination by the hindsight bias; knowledge about the later outcome may alter the neutral-practitioners' judgment about the propriety of earlier responses. One function of neutral-practitioners is to help define the envelope of appropriate responses given the information available to the practitioner at each point in the incident. Another function is to capture the real dilemmas, goal conflicts, and tradeoffs present in the actual workplace. In other words, the purpose is capture the ways that formal policies and procedures underspecify the demands of the field of practice.

An example occurred in regard to the Strasbourg aircraft crash (Monnier, 1992). Mode error in pilot interaction with cockpit automation seems to have been a contributor to this accident. Following the accident several people in the aviation industry noted a variety of precursor incidents for the crash where similar mode errors had occurred, although the incidents did not evolve as far towards negative consequences. These data provide us with information about what other similar practitioners have done, or would have done, when embedded in the context of commercial air transport. It indicates that a systemic vulnerability existed based on the design, rather than a simple case of a "human error."

Our research, and that of others, is based on the development of neutral-practitioner criteria for actions in complex systems. This method involves comparing actions that were taken by individuals to those of other similar practitioners placed in a similar situation. Note that this is a strong criterion for comparison and it requires that the evaluators possess or gather the same sort of expertise and experience as was employed during the incident. It does not rely on comparing practitioner behaviors with theory, rules, or policies. It is particularly effective for situations where the real demands of the system are poorly understood and where the pace of system activity level is fast or can cascade (i.e., in large, complex systems).

ERROR ANALYSIS AS CAUSAL JUDGMENT

Error and accident analysis is one case where people – lay people, scientists, engineers, managers, or regulators – make causal judgments or attributions. Causal attribution is a psychological and social judgment process that involves isolating one factor from among many contributing factors as a "cause" for the event to be explained. Strictly speaking, there are almost always several necessary and sufficient conditions for an event. But people distinguish among these necessary and sufficient conditions focusing on some as causes and relegating others to a background status as enabling conditions. In part, what is perceived as cause or enabling condition will depend on the context or causal background adopted (see Hart and Honore, 1959; also see Cheng and Novick, 1992). Consider a classic example used to illustrate this point. Oxygen is typically considered an enabling condition in an accident involving fire, as in the case of a dropped cigarette. However, people would generally consider oxygen as a cause if a fire broke out in a laboratory where oxygen was deliberately excluded as part of an experiment.

Current models of causal attribution processes hold that people attempt to explain the difference between the event in question and some contrasting case (or set of cases). Rather than explaining an event per se, one explains why the event occurs in the target case and not in some counterfactual contrast case (Hilton, 1990). Some relevant factors for establishing a causal background or contrast case are the dimensions originally proposed by Kelley (1973): "consensus, distinctiveness, and consistency". Consensus refers to the agreement between the responses of other people and the response of a particular person regarding a particular stimulus on a particular occasion; distinctiveness refers to the disagreement between the particular person's responses to some particular stimulus and other stimuli on the particular occasion; and consistency refers to the agreement between the way a particular person responds to a particular stimulus on different occasions (Cheng and Novick, 1992). The critical point is that there are degrees of freedom in how an event, such as an accident, is explained, and the explanation chosen depends, in part on the contrasting case or cases adopted. Thus, in a neutral-practitioner approach, the investigator tries to obtain data on different kinds of contrast cases, each of which may throw into relief different aspects of the dynamics of the incident in question.

Note that interactional or contrast case models of causal attribution help us to understand the diversity of approaches and attitudes towards "human error" and disasters. If someone asks another person why a particular incident occurred and if the shared background between these people is that "causes" of accidents are generally major equipment failures, environmental stresses, or misoperation, then it becomes sensible to respond that the incident was due to human error. If one asks why did a particular incident occur, when the shared background concerns identifying who is financially responsible (e.g., a legal perspective), then it becomes sensible to expect an answer that specifies the person or organization that erred. If questioner and respondent only appear to have a shared background (because both use the words "human error") when they, in fact, have different frames of reference for the question, then it is not surprising to find confusion.

In some sense, one could see the research of the 1980s on error as framing a different background for the question: Why did this incident occur? The causal background for the researchers involved in this intensive and cross-disciplinary examination of error and disaster was: How do we develop higher reliability complex human-machine systems? This causal background helped to point these researchers towards system-level factors in the management and design of the complex processes. In addition, when this question is posed by social and behavioral scientists, they (not so surprisingly) find socio-technical contributors (as opposed to reliability engineers who pointed to a different set of factors; Hollnagel, 1993). The benefit of the socio-technical background as a frame of reference for causal attribution is that it heightens our ability to go beyond the attribution of "human error" in analysis of risk and in measures to enhance safety. It seems to us that psychological processes of causal attribution apply as well to researchers on human error as they do to non-behavioral scientists. One could imagine a corollary for accident investigators to William James' Psychologist's Fallacy in which psychologists suppose that they are immune from the psychological processes that they study (Woods and Hollnagel, 2006).

The background for a neutral-practitioner approach to analyzing cognitive process and error comes from the local rationality assumption; that is, people do reasonable things, given their knowledge, objectives, point of view, and limited resources. However, an accident is by definition unintentional – people do not intend to act in ways that produce negative consequences (excepting sabotage). Error analysis traces the problem-solving process to identify points at which limited knowledge and processing lead to breakdowns. Process-tracing methods are used to map out how the incident unfolded over time, what the available cues were, which cues were actually noticed by participants, and how they were interpreted. Process tracing attempts to understand why the particular decisions/actions were taken; that is, how did it "make sense" to the practitioners embedded in the situation (Woods, 1993; Woods and Hollnagel, 2006; Dekker, 2006).

The relativistic notion of causal attribution suggests that we should seek out and rely on a broad set of contrast cases in explaining the sequence of events that led to an outcome. We explain why the practitioners did what they did by suggesting how that behavior could have been locally rational. To do this we need to understand behavior in the case in question relative to a variety of different contrast cases – what other practitioners have done in the situation or in similar situations. What we should not do, particularly when there is a demand to hold people accountable for their actions, is rely on putatively objective external evaluations of human performance such as those of court cases or other formal hearings. Such processes in fact institutionalize and legitimate the hindsight bias in the evaluation of human performance, easily leading to blame and a focus on individual actors at the expense of a system view.

14
ERROR AS INFORMATION

❖

One of the themes reverberating throughout this book is that human error represents a symptom rather than a cause. In this view error is, in part, a form of information about the functioning of the system in which those people are embedded (Rasmussen, 1986). We can use the idea of error-as-information to go behind the label human error and learn about how to improve human-machine, socio-technical systems. Lanir (1986) has developed a framework that captures how organizations can react to disaster.

A FUNDAMENTAL SURPRISE

On March 28, 1979, the U.S. nuclear industry and technologists were rocked by the Three Mile Island accident (TMI). The consternation that resulted was due to more than the fact that it was the worst nuclear accident up to that time or the radiological consequences per se. Rather, the accident is a case of what Lanir (1986) terms fundamental surprise. A fundamental surprise, in contrast to a situational surprise, is a sudden revelation of the incompatibility between one's self-perception and his environmental reality. Examples include the launch of Sputnik for the U.S., and the Yom Kippur war for Israel.

Perhaps the best way to grasp Lanir's concept of fundamental surprise is through an apocryphal story about Noah Webster, the well-known lexicographer (from Lanir, 1986). Lanir tells the story and then explains the concept this way:

CASE 14.1   APOCRYPHAL STORY ABOUT NOAH WEBSTER

One day, he arrived home unexpectedly to find his wife in the arms of his servant. 'You surprised me,' said his wife. 'And you have astonished me,' responded Webster. Webster's precise choice of words captured an important difference between his situation and that of his wife.
One difference between surprise and astonishment is the different level

of intensity associated with the two: astonishment is more powerful and extensive than surprise. Indeed, Mr. Webster's situation possesses an element of shock. His image of himself and his relations with his wife were suddenly and blatantly proven false. This was not the case for Mrs. Webster who, although surprised by the incident, still could maintain her image of herself, her environment, her husband, and the relations between them. Indeed, even if Mrs. Webster had taken all the steps she viewed as necessary to prevent the incident, she had to assume that there was some possibility of her unfaithfulness eventually being revealed. For Mrs. Webster, the failure was due to an external factor. Although she was uncertain about the external environment she was not uncertain about herself.

In contrast, Mr. Webster's astonishment revealed unrecognized uncertainty extending far beyond his wife, his servant, or other external factors. For him, comprehending the event's significance required a holistic re-examination of his self-perceptions in relation to his environment. Although this surprise offered Mr. Webster a unique opportunity for self-awareness, it came at the price of refuting his deepest beliefs.

A second distinction between surprise and astonishment lies in one's ability to define in advance the issues for which one must be alert. Surprises relate to specific events, locations, and time frames. Their demarcations are clear. Therefore, it is possible, in principle, to design early-warning systems to prevent them. In contrast, events providing astonishment affect broad scopes and poorly demonstrated issues. Mr. Webster's shocking incident revealed only the 'tip of an iceberg.'

Another distinction concerns the value of information. Mrs. Webster lacked one item of information which, had she had it in advance, would have allowed preventing her surprise: the information that her husband would return early that day. No single piece of information could have prevented Mr. Webster's astonishment. In most cases, the critical incident is preceded by precursors from which an outside observer could have deduced the state of the couple's relations. Such observers should be less prone to the tendency to interpret information in ways that suit their own world view, belittling or even ignoring the diagnostic value of information that contradicts it.

A fourth distinction between fundamental surprise and astonishment is in the ability to learn from the event. For Mrs. Webster, the learning process is simple and direct. Her early warning mechanisms were ineffective. If given a second chance, she might install a mechanism to reduce the possibility of being caught in a similar situational surprise.

Mr. Webster might attempt an explanation that would enable him to comprehend it without having to undergo the painful process of acknowledging and alerting a flawed world view. For example, he might blame the servant for 'attacking his innocent wife.' If it were established that the servant was not primarily at fault, he might explain the incident

> as an insignificant, momentary lapse on his wife's behalf. In more general terms, we may say that Mr. Webster's tendency to seek external, incidental reasons reflects the human tendency to behave as though astonishment is merely a surprise and, thus, avoid recognition of the need to experience painful 'self' learning. Lanir refers to Mrs. Webster's type of sudden discovery as a 'situational surprise' and Mr. Webster's sudden revelation of the incompatibility of his self-perception with this environmental reality as a 'fundamental surprise."

The TMI accident was more than an unexpected progression of faults; it was more than a situation planned for but handled inadequately; it was more than a situation whose plan had proved inadequate. The TMI accident constituted a fundamental surprise in that it revealed a basic incompatibility between the nuclear power industry's view of itself and reality. Prior to TMI, the industry could and did think of nuclear power as a purely technical system where all problems were in the form of some engineering technical area or areas, and the solutions to these problems lay in those engineering disciplines. TMI graphically revealed the inadequacy of that world view because the failures were in the socio-technical system and not due to pure technical factors (a single equipment or mechanical flaw) or to a purely human failure (gross incompetence or deliberate failures).

Prior to TMI, the pre-planning for emergencies consisted of considering large equipment failures; however, it did not consider a compounding series of small failures interacting with inappropriate human assessments of the situation and therefore erroneous actions. Prior to TMI, risk analysis also focused on large machine failures, not on the concatenation of several small failures, both machine and human. The kind of interaction between human and technical factors that actually occurred was inconceivable to the nuclear industry as a whole prior to TMI.

The post-TMI nuclear industry struggled to cope with, and adjust to, the revelations of TMI. The process of adjustment involved the phases associated with fundamental surprise described by Lanir:

○ First, the surprise event itself occurs.
○ Second, reaction spills over the boundaries of the event itself to include issues that have little to do with the triggering event – crises.
○ Third, these crises provide the opportunity for fundamental learning which, in turn, produces practical changes in the world in question.
○ Finally, the changes are absorbed and a new equilibrium is reached.

The immediate investigations of the TMI accident focused heavily on the mutual interaction between technical systems and people. The proposed changes that resulted from these investigations addressed the basic character of the joint human-machine system. These included providing new kinds of representations of the state of the plant, restructuring the guidance for board operators on how to handle abnormal conditions, and restructuring the organization of people in various facilities and their roles in handling different problems created by accidents.

However, in the process of carrying through on these and other "lessons learned", the U.S. nuclear industry shifted direction and treated the accident as if it was nothing more than a situational surprise. They began to focus on localized and purely technological solutions – what could be termed the *fundamental surprise error*, after Lanir's analysis (cf. Reason, 1990). This occurred despite the fact that the revelations of TMI continued to re-appear in other major incidents in the U.S. nuclear industry (e.g., the Davis-Besse nuclear power plant incident, see U.S. NRC, 1985) as well as in other risky technological worlds. While the post-TMI changes clearly have improved aspects of the socio-technical system through such things as new sensors, new analyses of possible accident conditions, new guidance on how to respond to certain accident conditions, and changes in emergency notification procedures, the basic socio-technical system for operating plants and responding to failures did not change (Moray and Huey, 1988).

As this case illustrates, incidents and accidents are opportunities for learning and change. But learning from the fundamental surprise may be partial and ineffective. The fundamental surprise often is denied by those involved. They interpret or redefine the incident in terms of local and specific factors as if it were only a situational surprise. The narrower interpretation can lead to denial of any need to change or to attribution of the "cause" to local factors with well-bounded responses – the "fundamental surprise error." In general, the fundamental surprise error is re-interpreting a fundamental surprise as merely a situational surprise which then requires no response or only a limited response. For the context of complex system failures, research results indicate that a specific version of this error is re-interpreting a human-machine system breakdown as being due to purely human factors. The label "human error" is a good example of a narrow interpretation that avoids confronting the challenges raised by the fundamental surprise. The fundamental surprise is that one must look for reliability in the larger system of interacting people and machines and the interactions between the layers of the system represented by the blunt and sharp ends (cf. recall the examples of human-machine system failures re-interpreted as simply "human error" cited earlier in this chapter). If the source of the incident is "human error," then only local responses are needed which do not require deep changes to the larger organization or system. Curing "human error" in this "local" sense only requires sanctions against the individuals involved, injunctions to try harder or follow the procedures more carefully, or some remedial adjustments in the training programs. Even more comfortable for the technologist is the thought that "human error" indicates that the people in the system are an unreliable component. This leads to the idea that "just a little more technology will be enough" (Woods, 1991) – that purely technological responses without consideration of human-machine systems or larger organizational factors can produce high reliability organizations. As a result of these rationalizations, the opportunity to learn from the fundamental surprise is lost.

As in the case of TMI, disasters in a variety of industries have been and continue to be unforeseen (Woods, 2005; Woods, 2009). As in the case of TMI, these accidents point to the interaction of people, technology, and the larger organization in which practitioners at the sharp end are embedded (Reason, 1997). The Thomas St. network failure challenges the larger organization and management systems for telecommunications (FCC, 1991); the Strasbourg and Bangalore crashes (Monnier, 1992; Lenorovitz, 1990) point to the human-machine cognitive system and the problems that can arise in coordination between people and automatic systems with many interacting modes. Before the fact, the accidents are largely inconceivable to the engineering and technological communities. As a result, Wagenaar and Groeneweg (1987) and Lanir (1986) have termed these accidents as "impossible", in the sense that the event is outside the closed world of a purely technical language of description. The challenge of fundamental surprise is to acknowledge these "impossible" events when they occur and to use them as sources of information for expanding the language of description. For us, the challenge is to expand the language of description to include systems of intertwined people and machines as in the joint or distributed cognitive system language used in Part III.

DISTANCING THROUGH DIFFERENCING

In the end, though, learning in the aftermath of incidents and accidents is extraordinarily difficult because of the complexity of modern systems. Layers of technical complexity hide the significance of subtle human performance factors. Awareness of hazard and the consequences of overt failure lead to the deployment of (usually successful) strategies and defenses against failure. These efforts create a setting where overt failures only occur when multiple small faults combine. The combination of multiple contributors and hindsight bias makes it easy for reviewers after the fact to identify an individual, group or organization as a culprit and stop. These characteristics of complex systems tend to hide the real characteristics of systems that lead to failures (Hollnagel et al., 2006).

When an organization experiences an incident, there are real, tangible and sometimes tragic consequences associated with the event which create barriers to learning:

- negative consequences are emotional and distressing for all concerned,
- failure generates pressure from different stakeholders to resolve the situation,
- a clear understandable cause and fix helps stakeholders move on from a tragedy, especially when they continue to use or participate in that system,
- managing financial responsibility for ameliorating the consequences and losses from the failure,
- desire for retribution from some stakeholders and processes of defense against punitive actions,
- confronting dissonance and changing concepts and ways of acting is painful and costly in non-economic senses.

The following case about how an organization struggled to recognize and overcome barriers to the learning process.

CASE 14.2   CHEMICAL FIRE

A chemical fire occurred during maintenance on a piece of process machinery in the clean room of a large, high-technology product manufacturing plant. The fire was detected and automatically extinguished by safety systems that shut off flow of reactants to the machine.

> The reactant involved in the fire was only one of many hazards associated with this expensive machine and the machine was only one of many arranged side by side in a long bay. Operation and maintenance of the machine also involved exposure or potential exposure of thermal, chemical, electrical, radio frequency, and mechanical hazards. Work in this environment was highly proceduralized and the plant was ISO-compliant. Both the risks of accident and the high value of the machine and its operation had generated elaborate formal procedures for maintenance and required two workers (buddy system) for most procedures on the machine.
>
> The manufacturer had an extensive safety program that required immediate and high-level responses to an incident such as this, even though no personal injury occurred and damage was limited to the machine involved. High-level management directed immediate investigations, including detailed debriefings of participants, reviews of corporate history for similar events, and a "root cause" analysis. Company policy required completion of this activity within a few days and formal, written notification of the event and related findings to all other manufacturing plants in the company. The cost of the incident may have been more than a million dollars.
>
> Two things prompted the company to engage outside consultants for a broader review of the accident and its consequences. First, search for prior similar events in the company files discovered a very similar accident at a manufacturing plant in another country earlier in the year. Second, one of the authors (RIC) recently had been in the plant to study the use of a different machine where operator "error" seemed prevalent but only with economic consequences. He had identified a systemic trap in this other case and provided some education about how complex systems fail a few weeks earlier. During that visit, he pointed out how other systemic factors could contribute to future incidents that threatened worker safety in addition to economic losses and suggested the need for broader investigations of future events.
>
> Following the incident two of the authors returned (RIC and DDW), visited the accident scene, and debriefed the participants in the event and those involved in its investigation. They studied operations involving the machine in which the fire occurred. They also examined the organizational response to this accident and to the prior fire.

The obstacles to learning from failure are nearly as complex and subtle as the circumstances that surround a failure itself. Because accidents always involve multiple contributors, the decision to focus on one or another of the set, and therefore what will be learned, is largely socially determined.

In the incident described in Case 14.2, the formal process of evaluating and responding to the event proceeded along a narrow path. The investigation concentrated on the machine itself, the procedures for maintenance, and the operators who performed the maintenance tasks. For example, they identified the fact the chemical reactant lines were clearly labeled outside the machine but not inside it where the maintenance took place. These local deficiencies were corrected quickly. In a sense, the accident was a 'normal' occurrence in the company; the event was regretted, undesirable, and costly, but essentially the sort of thing for which the company procedures had been designed and response teams created. The main findings of this formal, internal investigation were limited to these rather concrete, immediate, local items.

A broader review, conducted in part by outsiders, was based on using the specific incident as a wedge to explore the nature of technical work in context and how workers coped with the very significant hazards inherent in the manufacturing process. This analysis yielded a different set of findings regarding both narrow human engineering deficiencies and organizational issues. In addition to the relatively obvious human engineering deficiencies in the machine design discovered by the formal investigation, the event pointed to deeper issues that were relevant to other parts of the process and other potential events.

There were significant limitations in procedures and policies with respect to operations and maintenance of the machine. For example, although there were extensive procedural specifications contained in maintenance 'checklists', the workers had been called on to perform multiple procedures at the same time and had to develop their own task sequencing to manage the combination. Similarly, although the primary purpose of the buddy system was to increase safety by having one worker observe another to detect incipient failures, it was impossible to have an effective buddy system during critical parts of the procedures and parts of this maintenance activity. Some parts of the procedures were so complex that one person had to read the sequence from a computer screen while the other performed the steps. Other steps required the two individuals to stand on opposite sides of the machine to connect or remove equipment, making direct observation impossible.

Surprisingly, the formal process of investigating accidents in the company actually made deeper understanding of accidents and their sources more difficult. The requirement for immediate investigation and reporting contributed to pressure to reach closure quickly and led to a quick superficial study of the incident and its sources. The intense concern for "safety" had led the company to formally lodge responsibility for safety in a specific group of employees rather than the production and maintenance workers themselves. The need for safety as an abstract goal generated the need for these people as a separate entity within the company. These "safety people" had highly idealized views of the actual work environment, views uninformed by day-to-day contact with the realities of clean-room work conditions. These views allowed them to conceptualize the accident as flowing from the workers rather than the work situation. They were captivated in their investigation by physical characteristics of the workplace, especially those characteristics that suggested immediate, concrete interventions could be applied to "fix" the problems that they thought led to the accident.

In contrast, the operators regarded the incident investigation and proposed countermeasures as derived from views that were largely divorced from the realities of the workplace. They saw the "safety people" and their work as being irrelevant. They delighted in pointing out, for example, how few of them had any practical experience with working in the clean room. Privately, the workers said that production pressures were of paramount importance in the company. This view was communicated clearly to the workforce by multiple levels of management. Only after accidents, they noted, was safety regarded as a primary goal; during normal operations, safety was always a background issue, in contrast to the primary need to maintain high rates of production. The workers themselves internalized this view. There were significant incentives to provide directly to workers to obtain high production and they generally sought high levels of output to earn more money.

During the incident investigation, it was discovered that a very similar incident had occurred at another manufacturing plant in another country earlier in the year – a precursor event or rehearsal from the point of view of this manufacturing facility. Within the company, every incident, including the previous overseas fire, was communicated within the company to safety people and then on to other relevant parties. However, the formal report writing and dissemination about this previous incident had been slow and incomplete, relative to when the second event occurred. Part of the recommendations following from the second incident addressed faster production and circulation of reports (in effect, increasing the pressure to reach closure).

Interestingly, the relevant people at the plant knew all about the previous incident as soon as it had occurred through more informal communication channels. They had reviewed the incident, noted many features that were different from their plant (non-U.S. location, slightly different model of the same machine, different safety systems to contain fires). The safety people consciously classified the incident as irrelevant to the local setting, and they did not initiate any broader review of hazards in the local plant. Overall they decided the incident "couldn't happen here."

This is an instance of a discounting or distancing process whereby reviewers focus on differences, real and imagined, between the place, people, organization and circumstances where an incident happens and their own context. By focusing on the differences, they see few or no lessons for their own operation and practices.

Notice how speeding up formal notification does nothing to enhance what is learned and does nothing to prevent or mitigate discounting the relevance of the previous incident. The formal review and reports of these incidents focused on their unique features. This made it all the easier for audiences to emphasize what was different and thereby limit the opportunity to learn before they experienced their own incident.

It is important to stress that this was a company taking safety seriously. Within the industry it had an excellent safety record and invested heavily in safety. Its management was highly motivated and its relationships with workers were good, especially because of its strong economic performance that led to high wages and good working conditions. It recognized the need to make a corporate commitment to safety and to respond quickly to safety related events. Strong pressures to act quickly to "make it safe" provided incentives to respond immediately to each individual accident. But these demands in turn directed most of the attention after an accident towards specific countermeasures designed to prevent recurrence of that specific accident. This, in turn, led to the view that accidents were essentially isolated, local phenomena, without wider relevance or significance.

The management of the company was confronted with the fact that the handling of the overseas accident had not been effective in preventing the local one, despite their similarities. They were confronted by the effect of social processes working to isolate accidents and making them seem irrelevant to local operations. The prior fire overseas was noticed but regarded as irrelevant until after the local fire, when it suddenly became critically important information. It was not that the overseas fire was not communicated. Indeed it was observed by management and known even to the local operators. But these local workers regarded the overseas fire not as evidence of a type of hazard that existed in the local workplace but rather as evidence that workers at the other plant were not as skilled, as motivated and as careful as they were – after all, they were not Americans (the other plant was in a first world country). The consequence of this view was that no broader implications of the fire overseas were extracted locally after that event.

Interestingly (and ominously) this distancing through differencing that occurred in response to the external, overseas fire, was repeated internally after the local fire. Workers in the same plant, working in the same area in which the fire occurred but on a different shift, attributed the fire to lower skills of the workers on the other shift. Workers and managers of other parts of the manufacturing process also saw little relevance or potential to learn from the event. They regarded the workers to whom the accident happened as inattentive and unskilled. Not surprisingly, this meant that they saw the fire as largely irrelevant to their own work. After all, their reasoning went, the fire occurred because the workers to whom it happened were less careful than we are. Despite their beliefs, there was no evidence whatsoever that there were significant differences between workers on different shifts or in different countries (in fact, there was evidence that one of the workers involved was among the better skilled).

Contributing to this situation was, paradoxically, safety. Over a span of many years, the incidence of accidental fires with this particular chemical and in general had been reduced. But as a side effect of success, personnel's sensitivity to the hazard the chemical presented in the workplace was reduced as well. Interviews with experienced "old hands" in the industry indicated that such fires were once relatively common. New technical and procedural defenses against these events had reduced their frequency to the point that many operators had no personal experience with a fire. These "old hands" were almost entirely people now in management positions, far from the clean room floor itself. Those working with the hazardous materials were so young that they had no personal knowledge of these hazards, while those who did have experience were no longer involved in the day to day operations of the clean room.

In contrast with the formal investigation, the more extensive look into the accident that the outside researchers' visit provoked produced different findings. Discussion of the event prompted new observations from within the plant. Two examples may be given. One manager observed that the organization had extensive and refined policies for the handling of the flammable chemical delivery systems (tanks, pipes, valves) that stopped at the entrance to the machine. Different people, policies, and procedures applied to the delivery system. He made an argument for carrying these rules and policies through to the machine itself. This would have required more extensive (and expensive) preparation for maintenance on the machine than was currently the case, but would have eliminated the hazardous chemical from within the machine prior to beginning maintenance. Another engineer suggested that the absence of appropriate labeling on the machine involved with the accident should prompt a larger review of the labeling in all places where this chemical was used or transported.

These two instances are examples of using a specific accident to discover characteristics of the overall system. This kind of reasoning from the specific to the more general is a pronounced departure from the usual approach of narrowly looking for ways to prevent a very specific event from occurring or recurring. The chemical fire case reveals the pressures to discount or distance ourselves from incidents and accidents. In this organization, effective by almost all standards, managers, safety officers, and workers took a narrow view of the precursor event. By narrowing in on local, concrete, surface characteristics of the precursor event, the organization limited what could be learned.

EXTENDING OR ENHANCING THE LEARNING OPPORTUNITY

An important question is how to extend or enhance the window of opportunity for learning following incidents. The above case illustrates one general principle which could be put into action by organizations – do not discard other events because they appear on the surface to be dissimilar. At some level of analysis, all events are unique; while at other levels of analysis, they reveal common patterns.

Promoting means for organizations to look for and consider similarities between their own operation and the organization where an incident occurred could reduce the potential for distancing through differencing. This will require shifting analysis of the case from surface characteristics to deeper patterns and more abstract dimensions. Each kind of contributor to an event then can guide the search for similarities.

In the final analysis this organization in case 14.2 moved past the obstacle of distancing through differencing, and, as a result, the organization derived and shared a new lesson – safety is a *value* of an organization, not a commodity to be counted or a priority set among many other goals.

15
BALANCING ACCOUNTABILITY AND LEARNING

THE PRESSURE TO HOLD PEOPLE ACCOUNTABLE

When organizations make progress on safety by going behind the label "human error", it is mostly because they are able to do two things:

- Take a systems perspective: accidents are not caused by failures of individuals, but emerge from the conflux or alignment of multiple contributory system factors, each necessary and only jointly sufficient. The source of accidents is the system, not its component parts.
- Move beyond blame: blame focuses on the supposed defects of individual operators and denies the import of systemic contributions. In addition, blame has all kinds of negative side-effects. It typically leads to defensive posturing, obfuscation of information, protectionism, polarization, and mute reporting systems.

There can be pressures, however, to hold practitioners (or managers) involved in an incident or accident "accountable" even if that may hamper other people's willingness to voluntarily come forward with safety information in the future. Demands for accountability are not unusual or illegitimate, to be sure. Accountability is fundamental to any social relation. There is always an implicit or explicit expectation that we may be called upon to justify our beliefs and actions to others. The social-functionalist argument for accountability is that this expectation is mutual: as social beings we are locked into reciprocating relationships. Accountability, however, is not a unitary concept – even if this is what many stakeholders may think when aiming to improve people's performance under the banner of "holding them accountable". There are as many types of accountability as there are distinct relationships among people, and between people and organizations, and only highly specialized subtypes of accountability actually compel people to expend more cognitive effort. Expending greater effort, moreover, does not necessarily mean better task performance, as operators may become concerned more with limiting exposure and liability than with performing well (Lerner and Tetlock, 1999), something that can be observed in the decline of incident reporting with threats of prosecution. What is more, if accounting is perceived as illegitimate, for example intrusive, insulting or ignorant of real work, then any beneficial effects of accountability will vanish or backfire. Effects include a decline in motivation, excessive stress and attitude polarization, and the same effects can easily be seen in cases where practitioners or managers are "held accountable" by courts and other parties ignorant of the real trade-offs and dilemmas that make up actual operational work.

JUST CULTURE

The desire to balance learning from failure with appropriate accountability has motivated a number of safety-critical industries and organizations to develop guidance on a so-called "just culture." Of particular concern is the sustainability of learning from failure through incident reporting: if operators and others perceive that their reports will not be treated fairly or lead to negative consequences, the willingness to report will decline. The challenge for organizations, though, is that they want to know everything that happened, but cannot accept everything. The common assumption is that some behavior is inherently culpable, and should be treated as such. The public must be protected against intentional misbehavior or criminal acts, and the application of justice is a prime vehicle for this (e.g., Reason, 1997). As Marx (2001, p. 3) put it, "It is the balancing of the need to learn from our mistakes and the need to take disciplinary action that (needs to be addressed). Ultimately, it will help you answer the question: 'Where do you draw the disciplinary line?'"

Indeed, all proposals for building a just culture focus on drawing a clear line between acceptable and unacceptable behavior. For example, a just culture is one in which "front-line operators or others are not punished for actions, omissions or decisions taken by them that are commensurate with their experience and training, but where gross negligence, willful violations and destructive acts are not tolerated" (Eurocontrol, 2006). Such proposals emphasize the establishment of, and consensus around, some kind of separation between legitimate and illegitimate behavior: "in a just culture, staff can differentiate between acceptable and unacceptable acts" (Ferguson and Fakelmann, 2005, p. 34).

This, however, is largely a red herring. The issue of where the line is drawn is nowhere near as pressing as the question of who gets to draw it. An act is not an "error" in itself (this label, after all, is the result of social attribution), and an "error" is not culpable by itself either (that too, is the result of social attribution). Culpability does not inhere in the act. Whether something is judged culpable is the outcome of processes of interpretation and attribution that follow the act. Thus, to gauge whether behavior should fall on one side of the line or the other, people sometimes rely on culpability decision trees (e.g., Reason, 1997). Yet the questions it asks confirm the negotiability of the line rather than resolving its location:

○   *Were the actions and consequences as intended?* This evokes the judicial idea of a *mens rea* ("guilty mind"), and seems a simple enough question. Few people in safety-critical industries intend to inflict harm, though that does not prevent them from being prosecuted for their "errors" (under charges of manslaughter, for example, or general risk statutes that hail from road traffic laws on "endangering other people"; see for example Wilkinson, 1994). Also, what exactly is intent and how do you prove it? And who gets to prove this, using what kind of expertise?

○ *Did the person knowingly violate safe operating procedures?* People in all kinds of operational worlds knowingly violate safe operating procedures all the time. In fact, the choice can be as simple as either getting the job done or following all the applicable procedures. It is easy to show in hindsight which procedures would have been applicable and that they were available, workable and correct (says who, though?).

○ *Were there deficiencies in training or selection?* "Deficiencies" seems unproblematic but what is a deficiency from one angle can be perfectly normal or even above standard from another.

Questions such as the ones above may seem a good start, but they themselves cannot arbitrate between culpable or blameless behavior. Rather, they invoke new judgments and negotiations. This is true also for the very definition of negligence (a legal term, not a human performance concept):

> Negligence is conduct that falls below the standard required as normal in the community. It applies to a person who fails to use the reasonable level of skill expected of a person engaged in that particular activity, whether by omitting to do something that a prudent and reasonable person would do in the circumstances or by doing something that no prudent or reasonable person would have done in the circumstances. To raise a question of negligence, there needs to be a duty of care on the person, and harm must be caused by the negligent action. In other words, where there is a duty to exercise care, reasonable care must be taken to avoid acts or omissions which can reasonably be foreseen to be likely to cause harm to persons or property. If, as a result of a failure to act in this reasonably skillful way, harm/injury/damage is caused to a person or property, the person whose action caused the harm is negligent. (GAIN, 2004, p. 6)

There is no definition that captures the essential properties of "negligence." Instead, definitions such as the one above open a new array of questions and judgments. What is "normal standard"? How far is "below"? What is "reasonably skillful"? What is "reasonable care"? What is "prudent"? Was harm indeed "caused by the negligent action?" Of course, making such judgments is not impossible. But they remain judgments – made by somebody or some group in some time and place in the aftermath of an act – not objective features that stably inhabit the act itself.

ERROR AND CULPABILITY AS SOCIAL LABELS

Just as the properties of "human error" are not objective and independently existing, so does culpability arise out of our ways of seeing and putting things. What ends up being labeled as culpable does not inhere in the act or the person. It is constructed (or "constituted", as Christie put it) through the act of interrogation:

The world comes to us as we constitute it. Crime is thus a product of cultural, social and mental processes. For all acts, including those seen as unwanted, there are dozens of possible alternatives to their understanding: bad, mad, evil, misplaced honor, youth bravado, political heroism – or crime. The same acts can thus be met within several parallel systems as judicial, psychiatric, pedagogical, theological (Christie, 2004, p. 10):

> We would think that culpability, of all things, must make up some essence behind a number of possible descriptions of an act, especially if that act has a bad outcome. We seem to have great confidence that the various descriptions can be sorted out by the rational process of a peer-review or a hearing or a trial, that it will expose Christie's "psychiatric, pedagogical, theological" explanations (I had failure anxiety! I wasn't trained enough! It was the Lord's will!) as patently false. The application of reason will strip away the noise, the decoys, the excuses and arrive at the essential story: whether culpability lay behind the incident or not. And if culpable behavior turns out not make up the essence, then there will be no negative consequences.

But the same unwanted act can be construed to be a lot of things at the same time, depending on what questions we asked to begin with. Ask theological questions and we may see in an error the manifestation of evil, or the weakness of the flesh. Ask pedagogical questions and we may see in it the expression of underdeveloped skills. Ask judicial questions and we may begin to see a crime. Unwanted acts do not contain something culpable as their essence. We make it so, through the perspective we take, the questions we ask. As Christie argued, culpability is not an essence that we can discover behind the inconsistency and shifting nature of the world as it meets us. Culpability itself is that flux, that dynamism, that inconstancy, a negotiated arrangement, a tenuous, temporary stability achieved among shifting cultural, social, mental and political forces. Concluding that an unwanted act is culpable, is an accomplished project, a purely human achievement:

> Deviance is created by society ... social groups create deviance by making the rules whose infraction constitutes deviance and by applying those rules to particular persons and labeling them as outsiders. From this point of view, deviance is not a quality of the act the person commits, but rather a consequence of the application by others of rules and sanctions to an 'offender'. The deviant is the one to whom the label has successfully been applied; deviant behavior is behavior that people so label. (Becker, 1963, p. 9)

What counts as deviant or culpable is the result of processes of societal negotiation, of social construction. If an organization decides that a certain act constituted "negligence" or otherwise falls on the wrong side of the line, then this is the result of using a particular language and enacting a particular repertoire of post-conditions that turn the act into culpable behavior and the involved practitioner into an offender. Finding an act culpable, then, is a negotiated settlement onto one particular version of history. This version is not just produced for its own sake. Rather, it may serve a range of social functions, from emphasizing moral boundaries and enhancing solidarity, to sustaining subjugation or asymmetric power distribution within hierarchies, to protecting elite interests after an incident has exposed possibly expensive vulnerabilities in the system as a whole (Perrow, 1984), to mitigating public or internal apprehension about the system's ability to protect its safety-critical technologies against failure (Vaughan, 1996).

Who has the power to tell a story of performance in such a way – to use a particular rhetoric to describe it, ensuring that certain subsequent actions are legitimate or even possible (e.g., pursuing a single culprit), and others not – so as to, in effect, own the right to draw the line? This is a much more critical question than where the line goes, because that is anybody's guess. What is interesting is not whether some acts are so essentially negligent as to warrant more serious consequences. Instead, which processes or authorities does a society (or an organization) give the power to, to decide whether an act should be seen as negligent? Who enjoys the legitimacy to draw the line? The question for a just culture is not where to draw the line, but who gets to draw it.

ALTERNATIVE READINGS OF "ERROR"

People tend to believe that an "objective" account (one produced by the rational processes of a court, or an independent investigation of the incident) is superior in its accuracy because it is well-researched and not as tainted by interests or a particular, partisan perspective. These accounts, however, each represent only one tradition among a number of possible readings of an incident or unsafe act. They offer also just one language for describing and explaining an event, relative to a multitude of other possibilities. If we subscribe to one reading as true, it will blind us to alternative readings or framings that are frequently more constructive.

CASE 15.1 SURGICAL CRIMES?

Take as an example a British cardiothoracic surgeon who moved to New Zealand (Skegg, 1998). There, three patients died during or immediately after his operations, and he was charged with manslaughter. Not long before, a professional college had pointed to serious deficiencies in the surgeon's work and found that seven of his cases had been managed incompetently. The report found its way to the police, which subsequently investigated the cases. This in turn led to the criminal prosecution against the surgeon. Calling the surgical failures a crime is one possible interpretation of what went wrong and what should be done about it. Other ways are possible too, and not necessarily less valid:

- For example, we could see the three patients dying as an issue of cross-national transition: what are procedures for doctors moving to Australia or New Zealand and integrating them in local practice adequate?

> - And how are any cultural implications of practicing there systematically managed or monitored, if at all?
> - We could see these deaths as a problem of access control to the profession: do different countries have different standards for whom they would want as a surgeon, and who controls access, and how?
> - It could also be seen as a problem of training or proficiency-checking: do surgeons submit to regular and systematic follow-up of critical skills, as professional pilots do in a proficiency check every six months?
> - We could also see it as an organizational problem: there was a lack of quality control procedures at the hospital, and the surgeon testified having no regular junior staff to help with operations, but was made to work with only medical students instead.
> - Finally, we could interpret the problem as socio-political: what forces are behind the assignment of resources and oversight in care facilities outside the capital?

It may well be possible to write a compelling argument for each explanation of failure in the case above – each with a different repertoire of interpretations and countermeasures following from it. A crime gets punished away. Access and proficiency issues get controlled away. Training problems get educated away. Organizational issues get managed away. Political problems get elected or lobbied away. This also has different implications for what we mean by accountability. If we see an act as a crime, then accountability means blaming and punishing somebody for it. Accountability in that case is backward-looking, retributive. If, instead, we see the act as an indication of an organizational, operational, technical, educational or political issue, then accountability can become forward-looking. The question becomes: what should we do about the problem and who should bear responsibility for implementing those changes?

The point is not that one interpretation is right and all the others wrong. To even begin to grasp a phenomenon (such as an adverse surgical event in a hospital) we first have to accept the relevance and legitimacy of multiple, partially overlapping and often contradictory accounts. Because outside those, we have nothing. None of these accounts is inherently right and none is inherently wrong, but all of them are inherently limited. Telling the story from one angle necessarily excludes aspects from other angles. And all interpretations have different ramifications for what people and organizations think they should do to prevent recurrence, some more productive than others.

DISCRETIONARY SPACE FOR ACCOUNTABILITY

Telling multiple different stories of failure, however, can generate suspicions that operators simply want to blame the system. That they, as professionals (air traffic controllers, physicians, pilots) do not wish to be held accountable in the same way that others would be. Of course one message of this book is that we should look at the system in which people work, and improve it to the best of our ability. That, after all, is going behind the label "human error." But rather than presenting a false choice between blaming individuals or systems, we should explore the relationships and roles of individuals in systems. All safety-critical work is ultimately channeled through relationships between human beings (such as in medicine), or through direct contact of some people with the risky technology.

At this sharp end, there is almost always a discretionary space into which no system improvement can completely reach. This space can be filled only by an individual caregiving or technology-operating human. This is a final space in which a system really does leave people freedom of choice (to launch or not, to go to open surgery or not, to fire or not, to continue an approach or not). It is a space filled with ambiguity, uncertainty and moral choices. Systems cannot substitute the responsibility borne by individuals within that space. Individuals who work in those systems would not even want their responsibility to be taken away by the system entirely. The freedom (and the concomitant responsibility) there is probably what makes them and their work human, meaningful, a source of pride.

But systems can do two things. One is to be very clear about where that discretionary space begins and ends. Not giving practitioners sufficient authority to decide on courses of action (such as in many managed care systems), but demanding that they be held accountable for the consequences anyway, creates impossible and unfair double binds. Such double binds effectively shrink the discretionary space before action, but open it wide after any bad consequences of action become apparent (then it was suddenly the physician's or pilot's responsibility after all). Such vagueness or slipperiness of where the borders of the discretionary space lie is typical, but it is unfair and unreasonable.

The other thing for the system to decide is how to motivate people to carry out their responsibilities conscientiously inside that discretionary space. Is the source for that motivation going to be fear or empowerment? Anxiety or involvement? One common misconception is that "there has to be some fear that not doing one's job correctly could lead to prosecution." Indeed, prosecution presumes that the conscientious discharge of personal responsibility comes from fear of the consequences of not doing so. But neither civil litigation nor criminal prosecution work as a deterrent against human error. Instead, anxiety created by such accountability leads for example to defensive medicine, not high-quality care, and even to a greater likelihood of subsequent incidents (e.g., Dauer, 2004). The anxiety and stress generated by such accountability adds attentional burdens and distracts from conscientious discharge of the main safety-critical task (Lerner and Tetlock, 1999).

Rather than making people afraid, organizations could invest more in making people participants in change and improvement. Empowering people to affect their work conditions, to involve them in the outlines and content of that discretionary space, most actively promotes their willingness to shoulder their responsibilities inside of it.

CASE 15.2 PARALYTIC MISADMINISTRATION IN ANESTHESIA

Haavi Morreim (2004) recounts a case in which an anesthesiologist, during surgery, reached into a drawer that contained two vials, sitting side by side. Both vials had yellow labels and yellow caps. One, however, had a paralytic agent, and the other a reversal agent to be used later, when paralysis was no longer needed. At the beginning of the procedure, the anesthesiologist administered the paralyzing agent, as per intention. But toward the end, he grabbed the wrong vial, administering additional paralytic instead of its reversal agent. There was no bad outcome in this case. But when he discussed the event with his colleagues, it turned out that this had happened to them too, and that they were all quite aware of the enormous potential for confusion. All knew about the hazard, but none had spoken out about it. Anxiety about the consequences could be one explanation. There could have been a climate in which people were reluctant to contribute to improvements in their work, because of fear of the consequences of flagging their own potential errors. Perhaps, if they were to report their own syringe-swap near-miss, people felt at they could be sanctioned for being involved in an incident that could have harmed a patient. Do we think we can prevent anesthesiologists from grabbing a wrong vial by making them afraid of the consequences if they do? It is likely that anesthesiologists are sufficiently anxious about the consequences (for the patient) already. So should we not rather prevent them from grabbing a wrong vial by inviting them to come forward with information about that vulnerability, and giving the organization an opportunity to help do something structural about the problem?

That said, the problem of syringe swaps in anesthesia is well known in the anesthesia safety literature, but it also has proven to be impervious to quick solutions (see Cooper et al., 1978; Sandnes et al., 2008).

BLAME-FREE IS NOT ACCOUNTABILITY-FREE

Equating blame-free systems with an absence of personal accountability is inaccurate. Blame-free means blame-free, not accountability-free. The question is not whether practitioners want to skirt personal accountability. Few practitioners do. The question is whether we can meaningfully wring such accountability out of practitioners by blaming them, suing them or putting them on trial. We should instead convince ourselves that we can create such accountability not by blaming people, but by getting people actively involved in the creation of a better system to work in. Most practitioners will relish such responsibility, just as most practitioners often despair at the lack of opportunity to really influence their workplace and its preconditions for the better.

Holding people accountable and blaming people are two quite different things. Blaming people may in fact make them less accountable: they will tell fewer accounts, they may feel less compelled to have their voice heard, to participate in improvement efforts. Blame-free or no-fault systems are not accountability-free systems. On the contrary: such systems want to open up the ability for people to hold their account, so that everybody can respond and take responsibility for doing something about the problem. This also has different implications for what we mean by accountability. If we see an act as a crime, then accountability means blaming and punishing somebody for it. Accountability in that case is backward-looking, retributive. If, instead, we see the act as an indication of an organizational, operational, technical, educational or political issue, then accountability can become forward-looking (Sharpe, 2003). The question becomes what should we do about the problem and who should bear responsibility for implementing those changes. This, however, can get difficult very quickly. Lessons of an accident can get converted from an opportunity for a fundamental revision of assumptions about how the system works, to a mere local hiccup in an otherwise smooth operation (which can be taken care of by removing or punishing a few "bad apples").

BUILDING A JUST CULTURE

"What is just?" ask colleagues in the aftermath of an incident that happened to one of them. "How do we protect ourselves against disproportionate responses?" they add. "What is wise?" ask the supervisors. "What do people – other employees, customers, the public – expect me to do?" ask managers. And then other parties (e.g., prosecutors) ask, "Should we get involved?" The confusion about how to respond justly and still maintain a sense of organizational cohesion, loyalty and safety can be considerable.

At the same time, many organizations (whether they know it or not) seem to settle on pragmatic solutions that at least allow them to regain some balance in the wake of a difficult incident. When you look at these "solutions" a little more closely, you can see that they really boil down to answers to three central questions that need to dealt with in the process of building a just culture (Dekker, 2007):

- Who in the organization or society gets to draw the line between practitioners' acceptable and unacceptable behavior?
- What and where should the role of domain expertise be in judging whether behavior is acceptable or unacceptable?
- How protected against judicial interference are safety data (either the safety data from incidents inside of the organization or the safety data that come from formal accident investigations)?

The differences in the directions that countries or organizations or professions are taking towards just cultures come down to variations in the answers to these three questions. Some work very well, in some contexts, others less so. Also, the list of solutions is far from exhaustive, but it could inspire others to think more critically about where they or their organization may have settled (and whether that is good or bad).

In general, though, we can say this for the three questions. On the first question, the more a society, industry, profession or organization has made clear, agreed arrangements about who gets to draw the line, the more predictable the managerial or judicial consequences of an occurrence are likely to be. That is, practitioners may suffer less anxiety and uncertainty about what may happen in the wake of an occurrence, as arrangements have been agreed on and are in place.

On the second question, the greater the role of domain expertise in drawing the line, the less practitioners and organizations may be likely to get exposed to unfair or inappropriate judicial proceedings. That said, there is actually no research that suggests that domain experts automatically prevent the biases of hindsight slipping into their judgments of past performance. Hindsight is too pervasive a bias. It takes active reconstructive work, for everyone, to even begin to circumvent its effects. Also, domain experts may have other biases that work against their ability to fairly judge the quality of another expert's performance. There is, for example, the issue of psychological defense: if experts were to affirm that the potential for failure is baked into their activity and not unique to the practitioner who happened to inherit that potential, then this makes them vulnerable too.

On the third question, the better protected safety data is from judicial interference, the more likely practitioners could feel free to report. The protection of this safety data is connected, of course, to how the country or profession deals with the first and second question. For example, countries or professions that do protect safety data typically have escape clauses, so that the judiciary can gain access "when crimes are committed," or in "justified cases when duly warranted," or "for gross negligence and acts sanctioned by the criminal code." It is very important to make clear who gets to decide what counts as a "crime", or "duly warranted" or "gross negligence", because uncertainty (or the likelihood of non-experts making that judgment) can once again hamper practitioners' confidence in the system and their willingness to report or disclose.

16

SUMMING UP: HOW TO GO BEHIND THE LABEL "HUMAN ERROR"

" So what is human error? Can you give us a definition, or, better still, a taxonomy?" Throughout the history of safety, stakeholder groups have asked researchers for definitions and taxonomies of human error that are supposedly grounded in science. The questions are animated by a common concern: Each organization or industry feels that their progress on safety depends on having a firm definition of human error. Each group seems to believe that such a definition will enable creation of a scorecard that will allow them to gauge where organizations or industries stand in terms of being safe.

But each organization's search for the definition quickly becomes mired in complexity and terms of reference. Candidate definitions appear too specific for particular areas of operations, or too vague if they are broad enough to cover a wider range of activities. Committees are sometimes tasked with resolving the difficulties, but these produce ad hoc mixed-bag collections that only have a pretense of scientific standing. The definitions offered involve arbitrary and subjective methods of assigning events to categories. The resulting counts and extrapolations seem open to endless reassessment and debate. Beginning with the question "what is error?" misleads organizations into a thicket of difficulties where answers seem always just around the corner but never actually come into view.

The label "error" is used inconsistently in everyday conversations about safety and accidents. The term is used in at least three ways, often without stakeholders even knowing it:

- **Sense 1 – error as the cause of failure**: 'This event was due to human error.' The assumption is that error is some basic category or type of human behavior that precedes and generates a failure. It leads to variations on the myth that safety is protecting the system and stakeholders from erratic, unreliable people.
- **Sense 2 – error as the failure itself**, that is, the consequences that flow from an event: 'The transplant mix-up was an error.' In this sense the term "error" simply asserts that the outcome was bad producing negative consequences (e.g., injuries to a patient).

○ **Sense 3 – error as a process**, or more precisely, departures from the "good" process. Here, the sense of error is of deviation from a standard, that is a model of what is good practice. However, the enduring difficulty is that there are different models of what the process is that should be followed: for example, what standard is applicable, how standards should be described, and what it means when deviations from the standards do not result in bad outcomes. Depending on the model adopted, very different views of what is error result.

While you might think that it would always be clear from the context which of these senses people mean when they talk about error, in practice the senses are often confused with each other. Even worse, people sometimes slip from one sense to another without being aware that they are doing so.

As this book articulates, the search for definitions and taxonomies of error is not the first step on the journey toward safety; it is not even a useful step, only a dead end. Instead, the research on how individuals and groups cope with complexity and conflict in real-world settings has produced a number of insights about definitions of "error."

The first is that defining error-as-cause (Sense 1) blocks learning by hiding the lawful factors that affect human and system performance. This is the critical observation that gave rise to the idea that errors were heterogeneous and not directly comparable events that could be counted and tabulated. The standard way we say this today is that the label error should be the starting point of study and investigation, not the ending point.

Of course, it is tempting to stop the analysis of an adverse event when we encounter a person in the chain of events. Continuing the analysis through individuals requires workable models-cognition of individuals and of coordinated activity between individuals. It turns out to be quite hard to decide where to halt the causal analysis of a surprising event. Although there are theoretical issues involved in this stopping-rule problem, the decision about when to stop most often reflects our roles as stakeholders and as participants in the system. We stop when we think we have a good enough understanding and this understanding is, not surprisingly, when we have identified human error as the source of the failure.

The idea of error-as-cause also fails and misleads because it trivializes expert human performance. Error-as-cause leaves us with human performance divided in two: acts that are errors and acts that are non-errors. But this distinction evaporates in the face of any serious look at human performance (see for example, Klein, 1998). What we find is that the sources of successful operation of systems under one set of conditions can be what we label errors after failure occurs. Instead of finding error and non-error, when we look deeply into human systems at work we find that the behaviors there closely match the incentives, opportunities, and demands that are present in the workplace. Rather than being a distinct class of behavior, we find the natural laws that influence human systems are always at work, sometimes producing good outcomes and sometimes producing bad ones. Trying to separate error from non-error makes it harder to see these systemic factors.

Second, defining error-as-consequences (Sense 2) is redundant and confusing. Much of the time, the word "error" is used to refer to harm – generally preventable harm. This sort of definition is almost a tautology: it simply involves renaming preventable harm as error. But there are a host of assumptions that are packed into "preventable" and these are almost never made explicit. We are not interested in harm itself but, rather, how harm comes to be. The idea that something is preventable incorporates a complete (albeit fuzzy) model of how accidents happen, what factors contribute to them, and what sorts of countermeasures would be productive. But closer examination of "preventable" events shows that their preventability is largely a matter of wishing that things were other than they were.

To use "error" as a synonym for harm gives the appearance of progress where there is none. It would be better if we simply were clear in our use of language and referred to these cases in terms of the kind of harm or injuries. Confounding the label error with harm simply adds a huge amount of noise to the communication and learning process (for example, the label medication misadministration describes a kind of harm; the label medication error generates noise).

Third, defining error-as-deviation from a model of 'good' process (Sense 3) collides with the problem of multiple standards. The critical aspect of error-as-process-deviation is deciding how to determine what constitutes a deviation. Some have proposed normative models, for example Bayes Theorem, but these are rarely applicable to complex settings like health care, crisis management, or aviation and efforts to use this approach to assess human performance are misleading.

Some have argued that strict compliance with standard operating practices and procedures can be used to define deviation. However, it was quickly discovered that standard operating practices capture only a few elements of work and often prescribe practices that cannot actually be sustained in work worlds. In transportation systems, for example, where striking may be illegal, labor strife has sometimes led to workers adopt a "work-to-rule" strategy. By working exactly to rule, workers can readily make complex systems stop working. Attempts to make complete, exhaustive policies that apply to all cases creates or exacerbates double binds or to make it easy to attribute adverse events to 'human error' and stop. Expert performance is a lot more than following some set of pre-written guidance (Suchman, 1987).

Choosing among the many candidates for a standard changes what is seen as an error in fundamental ways. Using finer- or coarser-grain standards can give you a very wide range of error rates. In other words, by varying the standard seen as relevant, one can estimate hugely divergent 'error' rates. Some of the "standards" used in specific applications have been changed because too many errors were occurring or to prove that a new program was working. To describe something as a "standard" when it is capable of being changed in this way suggests that there is little that is standard about "standards". This slipperiness in what counts as a deviation can lead to a complete inversion of standardizing on good process: rather than describing what it is that people need to do to accomplish work successfully, we find ourselves relying on bad outcomes to specify what it is that we want workers not to do. Although often couched in positive language, policies, and procedures are often written and revised in just this way after accidents. Unfortunately, hindsight bias plays a major role in such activities.

Working towards meaningful standards as a means for assessing performance and defining error as deviations might be a long-term goal but it is fraught with hazard. To make standards work requires not only clear statements about how to accomplish work but clear guidance about how conflicts are to be handled. Specifying standards for performance for only part of the work to be done creates double binds that undermine expert-performance-creating conditions for failure. To use standards as a basis for evaluating performance deviations requires the continuous evaluation of performance against the standard rather than (as is often the case) simply after bad outcomes become apparent. One practical test of this is whether or not deviations from standards are actually detected and treated in the same way independent of the actual outcome.

To limit the damage from the multiple standards problem, all must carry forward in any tabulation the standard used to define deviations. This is absolutely essential! Saying some behavior was an error-as-process-deviation has no meaning without also specifying the standard used to define the deviation. There are three things to remember about the multiple standards problem:

○ First, the standard chosen is a kind of model of what it means to practice before outcome is known. A scientific analysis of human performance makes those models explicit and debatable. Without that background, any count is arbitrary.
○ Second, a judgment of error is not a piece of data which then can be tabulated with other like data; instead it is the end result of an analysis. Its interpretation rests on others being able to decompose and critique that analysis. The base data is the story of the particular episode – how multiple factors came together to produce that outcome. Effective systems of inquiry about safety begin with and continually refer back to these base stories of failure and of success to stimulate organizational learning.
○ Third, being explicit about the standard used is also essential to be able to critique, contrast, and combine results across events, studies, or settings. When these standards are dropped or hidden in the belief that error is an objective thing in the world, communication and learning collapse.

The research described in this book has shown that 'error' is an example of an essentially contestable concept. In fact, any benefit to the search for error only comes from the chronic struggle to define how different standards capture and fail to capture our current sense of what is expertise and our current model of the factors that make the difference between success and failure.

So, fourth, labeling an act as "error" marks the end of the social and psychological process of causal attribution. Research on how people actually apply the term "error" shows that "error" is a piece of data about reactions to failure, that is, it serves as a placeholder for a set of socially derived beliefs about how things happen. As stakeholders, our judgments after the fact about causality are used to explain surprising events. Thus, in practice, the study of error is nothing more or less than the study of the psychology and sociology of causal attribution. There are many regularities and biases – for example hindsight and oversimplifications – that determine how people judge causality. The heterogeneity and complexity of real-world work make these regularities and biases especially important: because the field of possible contributors includes so many items, biases may play an especially important role in determining which factors are deemed relevant.

This result is deeply unsettling for stakeholders because it points out that the use of the term "error" is less revealing about the performance of workers than it is about ourselves as evaluators. As researchers, advocates, managers, and regulators, we are at least as vulnerable to failure, susceptible to biases and oversimplifications, and prone to err as those other people at the sharp end. Fallibility has no bounds in a universe of multiple pressures, uncertainty, and finite resources.

Error is not a fixed objective, stable category or set of a categories for modeling the world. Instead, it arises from the interaction between the world and the people who create, run, and benefit (or suffer) from human systems for human purposes – a relationship between hazards in the world and our knowledge, our perceptions, and even our dread of the potential paths toward and forms of failure.

TEN STEPS

So if you feel you have a "human error" problem, don't think for a minute that you have said anything meaningful about the causes of your troubles, or that a better definition or taxonomy will finally help you get a better grasp of the problem, because you are looking in the wrong place, and starting from the wrong position. You don't have a problem with erratic, unreliable operators. You have an organizational problem, a technological one. You have to go *behind* the label human error to begin the process of learning, of improvement, of investing in safety. In the remainder of this concluding chapter, we walk through 10 of the most important steps distilled from the research base about how complex systems fail and how people contribute to safety. The 10 steps forward summarize general patterns about error and expertise, complexity, and learning. These 10 steps constitute a checklist for constructive responses when you see a window of opportunity to improve safety. Here they are:

1. Recognize that human error is an attribution.
2. Pursue second stories to find deeper, multiple contributors to failure.
3. Escape the hindsight bias.
4. Understand work as performed at the sharp end of the system.
5. Search for systemic vulnerabilities.
6. Study how practice creates safety.
7. Search for underlying patterns.
8. Examine how change will produce new vulnerabilities and paths to failure.
9. Use new technology to support and enhance human expertise.
10. Tame complexity.

1. RECOGNIZE THAT HUMAN ERROR IS AN ATTRIBUTION

"Human error" (by any other name: procedural violation, deficient management) is an attribution. It is not an objective fact that can be found by anybody with the right method or right way of looking at an incident. Given that human error is just an attribution, that it is just one way of saying what the cause was, just one way of telling a story about a dreadful event (a first story); it is entirely justified for us to ask why telling the story that way makes sense to the people listening.

Attributing failure to "human error" has many advantages. The story that results, the first story, is short and crisp. It is simple. It is comforting because it shows that we can quickly find causes. It helps people deal with anxiety about the basic safety of the systems on which we depend. It is also a cheap story because not much needs to be fixed: just get rid of or remediate the erratic operators in the system. It is a politically convenient story, because people can show that they are doing something about the problem to reassure the public.

But if that story of a failure is only one attribution of its cause, then other stories are possible too. We are entirely justified to ask who stands to gain from telling a first story and stopping there. Whose voice goes unheard as a result? What stories, different stories, with different attributions about the failure, get squelched?

The first story after celebrated accidents tells us nothing about the factors that influence human performance before the fact. Rather, the first story represents how we, with knowledge of outcome and as stakeholders, react to failures. Reactions to failure are driven by the consequences of failure for victims and other stakeholders. Reactions to failure are driven, too, by the costs of changes that would need to be made to satisfy stakeholders that the threats exposed by the failure are under control. This is a social and political process about how we attribute "cause" for dreadful and surprising breakdowns in systems that we depend on and that we expect to be safe.

So a story of human error is merely a starting point to get at what went wrong, not a conclusion about what went wrong. A first story that attributes the failure to error is an invitation to go deeper, to find those other voices, to find those other stories, and to discover a much deeper complexity of what makes systems risky or safe.

What is the consequence of error being the result of processes of attribution after the fact? The belief that there is an answer to the question – What is error? – is predicated on the notion that we can and should treat error as an objective property of the world and that we can search for errors, tabulate them, count them. This searching and counting is futile.

The relationship between error and safety is mirage-like. It is as if you find yourself in a desert, seeing safety glimmering somewhere in the far distance. To begin the journey, you feel we must gauge the distance to your goal in units of "error." Yet your presumption about the location of safety is illusory. Efforts to measure the distance to it are little more than measuring your distance from a mirage. The belief that estimates of this number are a necessary or even useful method of beginning an effort to improve safety is false.

The psychology of causal attribution reminds us that it is our own beliefs and misconceptions about failure and error that have combined to make the mirage appear where it does. The research on how human systems cope with complexity tells us that progress towards safety has more to do with the metaphorical earth underneath your feet than it does with that tantalizing image off in the distance. When you look down, you see people struggling to anticipate forms of, and paths toward, failure. You see them actively adapting to create and sustain failure-sensitive strategies, and working to maintain margins in the face of pressures to do more and do it quickly. Looking closely under your feet, you may begin to see:

- how workers and organizations are continually revising their approach to work in an effort to remain sensitive to the possibility for failure;
- how we and the workers are necessarily only partially aware of the current potential for failure;
- how change is creating new paths to failure and new demands on workers and how revising their understanding of these paths is an important aspect of work on safety;
- how the strategies for coping with these potential paths can be either strong and resilient or weak and brittle;
- how the culture of safety depends on remaining dynamically engaged in new assessments and avoiding stale, narrow, or static representations of risk and hazard;
- how overconfident nearly everyone is that they have already anticipated the types and mechanisms of failure, and how overconfident nearly everyone is that the strategies they have devised are effective and will remain so;
- how missing side effects of change is the most common form of failure for organizations and individuals; and
- how continual effort after success in a world of changing pressures and hazards is fundamental to creating safety.

But for you to see all that, you have to go behind the label human error. You have to relentlessly, tirelessly pursue the deeper, second story.

2. PURSUE SECOND STORIES

First stories, biased by knowledge of outcome, are overly simplified accounts of the apparent "cause" of the undesired outcome. The hindsight bias narrows and distorts our view of practice after the fact. As a result, there is premature closure on the set of contributors that lead to failure, the pressures and dilemmas that drive human performance are masked. A first story always obscures how people and organizations work to recognize and overcome hazards and make safety.

Stripped of all the context, first stories are appealing because they are easy to tell and they locate the important "cause" of failure in practitioners closest to the outcome. First stories appear in the press and usually drive the public, legal, political, and regulatory reactions to failure. Unfortunately, first stories simplify the dilemmas, complexities, and difficulties practitioners face and hide the multiple contributors and deeper patterns. The distorted view they offer leads to proposals for "solutions" that are weak or even counterproductive and blocks the ability of organizations to learn and improve.

First stories cause us to ask questions like: "How big is this safety problem?", "Why didn't someone notice it before?" and "Who is responsible for this state of affairs?" The calls to action based on first stories have followed a regular pattern:

- Demands for increasing the general awareness of the issue among the public, media, regulators, and practitioners ("we need a conference…");
- Calls for others to try harder or be more careful ("those people should be more vigilant about …")
- Insistence that real progress on safety can be made easily if some local limitation is overcome ("we can do a better job if only…");
- Calls for more extensive, more detailed, more frequent, and more complete, reporting of problems ("we need mandatory incident reporting systems with penalties for failure to report …"); and
- Calls for more technology to guard against erratic people ("we need computer order entry, bar coding, electronic medical records, and so on …").

First stories are not an explanation of failure. They merely represent, or give expression to, a reaction to failure that attributes the cause of accidents to narrow proximal factors, usually "human error." They appear to be attractive explanations for failure, but they lead to sterile or even counterproductive responses that limit learning and improvement (e.g., "we need to make it so costly for people that they will have to …").

When you see this process of telling first stories go on, the constructive response is very simple – in principle. Go beyond the first story to discover what lies behind the term "human error." Your role could be to help others develop the deeper second story. This is the most basic lesson from past research on how complex systems fail. When you pursue second stories, the system starts to look very different. You can begin to see how the system moves toward, but is usually blocked from, accidents. Through these deeper insights learning occurs and the process of improvement begins. Progress on safety begins with uncovering second stories.

3. ESCAPE FROM HINDSIGHT BIAS

Knowledge of outcome distorts our view of the nature of practice in predictable and systematic ways. With knowledge of outcome we simplify the dilemmas, complexities, and difficulties practitioners face and how they usually cope with these factors to produce success. The distorted view leads people to propose "solutions" that actually can be counterproductive if they degrade the flow of information that supports learning about systemic vulnerabilities and if they create new complexities to plague practice. In contrast, research-based approaches try to use various techniques to escape from hindsight bias. This is a crucial prerequisite for learning to occur.

4. UNDERSTAND THE WORK PERFORMED AT THE SHARP END OF THE SYSTEM

When you start to pursue the second story, the way you look at people working at the sharp end of a system changes dramatically. Instead of seeing them and their work as the instigators of trouble, as the sources of failure, you begin to see the sharp end as the place where many of the pressures and dilemmas of the entire system collect, where difficult situations are resolved on a daily basis. These are organizational, economic, human, and technological factors, that flow toward the sharp end and play out to create outcomes. Sharp-end practitioners who work in this setting face of a variety of difficulties, complexities, dilemmas, and tradeoffs, they are called on to achieve multiple, often conflicting, goals. Safety is created at the sharp end as practitioners interact with hazardous processes inherent in the field of activity, using the available tools and resources.

To really build and understand a second story, you have to look at more than just one incident or accident. You have to go more broadly than a single case to understand how practitioners at the sharp end function – the nature of technical work as experienced by the practitioner in context.

Ultimately, all efforts to improve safety will be translated into new demands, constraints, tools, or resources that appear at the sharp end. Improving safety depends on investing in resources that support practitioners meet the demands and overcome the inherent hazards in that setting. In other words, progress on safety depends on understanding how practitioners cope with the complexities of technical work.

When you shift your focus to technical work in context, you actually begin to wonder how people usually succeed. Ironically, understanding the sources of failure begins with understanding how practitioners create success and safety first; how they coordinate activities in ways that help them cope with the different kinds of complexities they experience. Interestingly, the fundamental insight that once launched the New Look was to see human performance at work as human adaptations directed to cope with complexity (Rasmussen, 1986; Woods and Hollnagel, 2006). Not understanding the messy details of what it means to practice in an operational setting, and being satisfied with only shallow, short forays into the real world of practitioners, carries big risks:

> "The potential cost of misunderstanding technical work" is the risk of setting policies whose actual effects are "not only unintended but sometimes so skewed that they exacerbate the problems they seek to resolve. Efforts to reduce 'error' misfire when they are predicated on a fundamental misunderstanding of the primary sources of failures in the field of practice [systemic vulnerabilities] and on misconceptions of what practitioners actually do." (Barley and Orr, 1997, p. 18)

Here are three ways to help you focus your efforts to understand technical work as it affects the potential for failure:

a. **Look for sources of success**. To understand failure, try to understand success in the face of complexities. Failures occur in situations that usually produce successful outcomes. In most cases, the system produces success despite opportunities to fail. To understand failure requires understanding how practitioners usually achieve success in the face of demands, difficulties, pressures, and dilemmas. Indeed, as Ernst Mach reminded us more than a century ago, success and failure flow from the same sources.

b. **Look for difficult problems**. To understand failure, look for difficult problems, and try to understand what makes them difficult. Patterson et al. (2010) provide the latest listing of facets of complexity. Aim to identify the factors that made certain situations more difficult to handle and explore the individual and team strategies used to handle these situations. As you begin to understand what made certain kinds of problems difficult, how expert strategies were tailored to these demands, and how other strategies were poor or brittle, you may begin to discover new ways to support and broaden the application of successful strategies.

c. **Avoid the Psychologist's Fallacy**. Understand the nature of practice from practitioners' point of view. It is easy to commit what William James called the Psychologist's Fallacy in 1890. Updated to today, this fallacy occurs when well-intentioned observers think that their distant view of the workplace captures the actual experience of those who perform technical work in context. Distant views can miss important aspects of the actual work situation and thus can miss critical factors that determine human performance in that field of practice. To avoid the danger of this fallacy, cognitive anthropologists use research techniques based on a practice-centered perspective. Researchers on human problem-solving and decision-making refer to the same concept with labels such as process-tracing and naturalistic decision-making (Klein, Orasanu and Calderwood, 1993).

It is important to distinguish clearly that doing technical work expertly is not the same thing as expert understanding of the basis for technical work. This means that practitioners' descriptions of how they accomplish their work often are biased and cannot be taken at face value. For example, there can be a significant gap between people's descriptions (or self-analysis) of how they do something and observations of what they actually do. Successful practice-centered demands a combination of the following three factors:

○ the view of practitioners in context,
○ technical knowledge in that area of practice, and
○ knowledge of general results/concepts about the various aspects of human performance that play out in that setting.

Since technical work in context is grounded in the details of the domain itself, it is also insufficient to be expert in human performance in general. Understanding technical work in context requires (1) in-depth appreciation of the pressures and dilemmas practitioners face and the resources and adaptations practitioners bring to bear to accomplish their goals, and also (2) the ability to step back and reflect on the deep structure of factors that influence human performance in that setting. Individual observers rarely possess all of the relevant skills so that progress on understanding technical work in context and the sources of safety inevitably requires interdisciplinary cooperation.

5. SEARCH FOR SYSTEMIC VULNERABILITIES

Through practice-centered observation and studies of technical work in context, safety is not found in a single person, device or department of an organization. Instead, safety is created and sometimes broken in systems, not individuals. The issue is finding systemic vulnerabilities, not flawed individuals. Safety is an emergent property of systems, not of their components.

Examining technical work in context with safety as your purpose, you will notice many hazards, complexities, gaps, tradeoffs, dilemmas, and points where failure is possible. You will also begin to see how the practice of operational people has evolved to cope with these kinds of complexities. After elucidating complexities and coping strategies, one can examine how these adaptations are limited, brittle, and vulnerable to breakdown under differing circumstances. Discovering these vulnerabilities and making them visible to the organization is crucial if we are anticipate future failures and institute change to head them off. Indeed, detection and recovery is critical to success. You should aim to understand how the system supports (or fails to support) detection of and recovery from failures.

Of course, this process of feedback, learning, and adaptation should go on continuously across all levels of an organization. As changes occur, some vulnerabilities decay while new paths to failure emerge. To track the shifting pattern requires getting information about the effects of change on sharp-end practice and about new kinds of incidents that begin to emerge. If the information is rich enough and fresh enough, it is possible to forecast future forms of failure, to share schemes to secure success in the face of changing vulnerabilities. Producing and widely sharing this sort of information may be one of the hallmarks of a culture of safety.

However, establishing a flow of information about systemic vulnerabilities is quite difficult because it is frightening to consider how all of us, as part of the system of interest, can fail. Repeatedly, research notes that blame and punishment will drive this critical information underground. Without a safety culture, systemic vulnerabilities become visible only after catastrophic accidents. In the aftermath of accidents, learning also is limited because the consequences provoke first stories, simplistic attributions, and shortsighted fixes.

Understanding the 'systems' part of safety involves understanding how the system itself learns about safety and responds to threats and opportunities. In organizational safety cultures, this activity is prominent, sustained, and highly valued. The learning processes must be tuned to the future to recognize and compensate for negative side effects of change and to monitor the changing landscape of potential paths to failure. It is critical to examine how the organization at different levels of analysis supports or fails to support the process of feedback, learning, and adaptation. In other words, find out how well the organization is learning how to learn. Safe organizations deliberately search for and learn about systemic vulnerabilities.

6. STUDY HOW PRACTICE CREATES SAFETY

Typically, reactions to failure assume the system is "safe" (or has been made safe) inherently and that overt failures are only the mark of an unreliable component. But what is irreducible is uncertainty about the future, change is always active, and resources are always finite. As a result, all systems confront inherent hazards, tradeoffs, and are vulnerable to failure. Second stories reveal how practice is organized to allow practitioners to create success in the face of threats. Individuals, teams, and organizations are aware of hazards and adapt their practices and tools to guard against or defuse these threats to safety. It is these efforts that "make safety." This view of the human role in safety has been a part of complex systems research since its origins, and encourages you to study how practitioners cope with hazards and resolve tradeoffs; how they mostly succeed, yet sometimes fail.

Adaptations by individuals, teams, and organizations that have worked well in the past can become limited or stale. This means that feedback about how well adaptations are working or about how the environment is changing is critical. Examining the weaknesses and strengths, costs and benefits of these adaptations points to the areas ripe for improvement. As a result, progress depends on studying how practice creates safety in the face of challenges, what it takes to be an expert in context.

7. SEARCH FOR UNDERLYING PATTERNS

In discussions of some particular episode of failure, or of some "hot button" safety issue, it is easy for commentators to examine only the surface characteristics of the area in question. Progress has come from going beyond the surface descriptions (the phenotypes of failures) to discover underlying patterns of systemic factors (Hollnagel, 1993).

Genotypes are concepts and models about how people, teams, and organizations coordinate information and activities to handle evolving situations and cope with the complexities of that work domain. These underlying patterns are not simply about knowledge of one area in a particular field of practice. Rather, they apply, test, and extend knowledge about how people contribute to safety and failure and how complex systems fail by addressing the factors at work in this particular setting. As a result, when we examine technical work, search for underlying patterns by contrasting sets of cases.

8. EXAMINE HOW ECONOMIC, ORGANIZATIONAL AND TECHNOLOGICAL CHANGE WILL PRODUCE NEW VULNERABILITIES AND PATHS TO FAILURE

As capabilities, tools, organizations, and economic pressures change, vulnerabilities to failure change as well. This means that the state of safety in any system always is dynamic, and that maintaining safety in that system is a matter of maintaining dynamic stability, not static stability. Systems exist in a changing world. The environment, organization, economics, capabilities, technology, management, and regulatory context all change over time. This backdrop of continuous systemic change ensures that hazards and how they are managed are constantly changing. Also, a basic pattern in complex systems is a drift toward failure as planned defenses erode in the face of production pressures and change. As a result, when we examine technical work in context, we need to understand how economic, organizational, and technological change can create new vulnerabilities in spite of or in addition to providing new benefits.

Research reveals that organizations that manage potentially hazardous technical operations remarkably successfully create safety by anticipating and planning for unexpected events and future surprises. These organizations did not take past success as a reason for confidence. Instead, they continued to invest in anticipating the changing potential for failure because of the deeply held understanding that their knowledge base was fragile in the face of the hazards inherent in their work and the changes omnipresent in their environment.

Under resource pressure, however, any safety benefits of change can get quickly sucked into increased productivity, which pushes the system back to the edge of the performance envelope. Most benefits of change, in other words, come in the form of increased productivity and efficiency and not in the form of a more resilient, robust, and therefore safer system (Rasmussen, 1986). Researchers in the field speak of this observation as the Law of Stretched Systems (Woods and Hollnagel, 2006):

> We are talking about a law of systems development, which is every system operates, always at its capacity. As soon as there is some improvement, some new technology, we stretch it. (Hirschhorn, 1997)

Change under resource and performance pressures tends to increase coupling, that is, the interconnections between parts and activities, in order to achieve greater efficiency and productivity. However, research has found that increasing coupling also increases operational complexity and increases the difficulty of the problems practitioners can face (Woods, 1988). Increasing the coupling between parts in a process changes how problems manifest, creating or increasing complexities such as more effects at a distance, more and faster cascades of effects and tighter goal conflicts. As a result, increased coupling between parts creates new cognitive and collaborative demands which contribute to new forms of failure.

Because all organizations are resource limited to one degree or another, you are probably naturally concerned with how to prioritize issues related to safety. Consider focusing your resources on anticipating how economic, organizational, and technological change could create new vulnerabilities and paths to failure. Armed with any knowledge produced by this focus, you can try to address or eliminate these new vulnerabilities at a time when intervention is less difficult and less expensive (because the system is already in the process of change). In addition, these points of change are at the same time opportunities to learn how the system actually functions.

9. USE NEW TECHNOLOGY TO SUPPORT AND ENHANCE HUMAN EXPERTISE

The notion that it is easy to get "substantial gains" through computerization or other forms of new technology is common in many fields. The implication is that new technology by itself reduces human error and minimizes the risk of system breakdown. Any difficulties that are raised about the computerization process or about the new technology become mere details to be worked out later.

But this idea, that a little more technology will be enough, has not turned out to be the case in practice. Those pesky details turn out to be critical in whether the technology creates new forms of failure. New technology can help and can hurt, often at the same time – depending on how the technology is used to support technical work in context. Basically, it is the underlying complexity of operations that contributes to the human performance problems. Improper computerization can simply exacerbate or create new forms of complexity to plague operations. The situation is complicated by the fact the new technology often has benefits at the same time that it creates new vulnerabilities. New technology cannot simply be thrown at a world of practice, as if it is merely one variable that can be controlled independently of all others. People and computers are not separate and independent, but are interwoven into a distributed system that performs cognitive work in context. Changing anything about that intricate relationship immediately changes that joint system's ability to create success or forestall failure.

The key to skillful as opposed to clumsy use of technological possibilities lies in understanding the factors that lead to expert and team performance and the factors that challenge expert and team performance. The irony is that once you understand the factors that contribute to expertise and to breakdown, you will probably understand how to use the powers of the computer to enhance expertise. On the one hand, new technology creates new dilemmas, demands, knowledge and memory requirements and new judgments. But, on the other hand, once the basis for human expertise and the threats to that expertise had been studied, technology can be an important means to enhance system performance. As a result, when you examine technical work, try once again to understand the sources of and challenges to expertise in context. This is crucial to guide the skillful, as opposed to clumsy use of technological possibilities.

10. TAME COMPLEXITY THROUGH NEW FORMS OF FEEDBACK

Failures represent breakdowns in adaptations directed at coping with complexity. Success relates to organizations, groups, and individuals who are skillful at recognizing the need to adapt in a changing, variable world and in developing ways to adapt plans to meet these changing conditions despite the risk of negative side effects.

Recovery before negative consequences occur, adapting plans to handle variations and surprise, recognizing side effects of change are all critical to high resilience in human and organizational performance. Yet all of these processes depend fundamentally on the ability to see the emerging effects of decisions, actions, policies. It depends on the kind of feedback a team or an organization makes available about itself and to itself, especially feedback about the future. In general, increasing complexity can be balanced with improved feedback. Improving feedback is a critical investment area for improving human performance and guarding against paths toward failure.

The constructive response to issues on safety is to study where and how to invest in better feedback. This is a complicated subject since better feedback is:

○ integrated to capture relationships and patterns, not simply a large set of available but disconnected data elements (like those gathered in many incident reporting or data monitoring systems);
○ event-based to capture change and sequence over multiple time-scales, not simply the current values on each data channel;

○ future-oriented to help organizational decision makers assess what could happen next, not simply what has happened;
○ context-sensitive and tuned to the interests and expectations of those looking at the data.

Feedback at all levels of the organization is critical. Remember, a basic pattern in complex systems is a drift toward failure as planned defenses erode in the face of production pressures, and as a result of changes that are not well-assessed for their impact on the cognitive work that goes on at the sharp end. Continuous organizational feedback is needed to support adaptation and learning processes. Paradoxically, feedback must be tuned to the future to detect the emergence of drift toward failure, to explore and compensate for negative side effects of change, and to monitor the changing landscape of potential paths to failure. To achieve this, you should help your organization develop and support mechanisms that create foresight about the constantly changing shape of the risks it faces.

CONCLUSION

Safety is not a commodity to be tabulated, it is a chronic value 'under your feet' that infuses all aspects of practice. People create safety under resource and performance pressure at all levels of socio-technical systems. They continually learn and adapt their activities in response to information about failure. Progress on safety does not come from hunting down those who err. Instead, progress on safety comes from finding out what lies behind attributions of error – the complexity of cognitive systems, the messiness of organizational life, and ultimately your own reactions, anxieties, hopes, and desires as a stakeholder and as a participant in human systems serving human purposes. Progress on safety comes from going behind the label human error, where you discover how workers and managers create safety, and where you find opportunities to help them do it even better.

REFERENCES

ACCIDENT/INCIDENT REPORTS

Air Accident Investigation and Aviation Safety Board (AAIASB) (2006). *Aircraft accident report (11/2006): Helios Airways flight HCY522, Boeing 737–31S at Grammatiko, Hellas on 14 August 2005*. Athens: Helenic Republic Ministry of Transport and Communications.
Aviation Week and Space Technology (1995). A 330 Test Order Included Study of Autopilot Behavior. 04/10/95, p. 60.
Aviation Week and Space Technology (1995). Toulouse A330 Flight Swiftly Turned Critical. 04/17/95, p. 44.
Columbia Accident Investigation Board (2003). *Report Volume 1, August 2003*. Washington, DC: U.S. Government Printing Office.
Cook, R.I. and Nemeth, C., (2007). The "Milrinone Event" November, 2006. Report on the Investigation of a Healthcare Adverse Event. Medical Event Data Collection and Analysis Service (MEDCAS). University of Chicago, Cognitive Technologies Laboratory.
Cook, R.I., Woods D.D., and Howie, M.B. (1992). Unintentional delivery of vasoactive drugs with an electromechanical infusion device. *Journal of Cardiothoracic and Vascular Anesthesia*, 6, 238–244.
Dutch Safety Board (1999). *Final report 97–75/A-26, PH-TKC Boeing 757, 24 December 1997 Amsterdam Airport Schiphol*. The Hague, NL: Author.
Eurocontrol Performance Review Commission (2006). *Report on legal and cultural issues in relation to ATM safety occurrence reporting in Europe: Outcome of a survey conducted by the Performance Review Unit in 2005–2006*. Brussels: Eurocontrol.
Federal Communications Commission (1991). *Report on the September 17, 1991 AT&T NY power outage at Thomas Street switching station and network disruption*. Washington DC: Common Carrier Bureau.
Fire Investigation Team (2000). Cerro Grande Prescribed Fire Investigative Report, National Interagency Fire Center, Boise ID. May 18, 2000.

Freund, P.R. and Sharar, S.R. (1990). Hyperthermia alert caused by unrecognized temperature monitor malfunction. *Journal of Clinical Monitoring*, 6, 257.

Johnston, A.M. (2006). Report of an investigation by the Inspector appointed by the Scottish Ministers for The Ionising Radiation (Medical Exposures) Regulations 2000. Unintended overexposure of patient Lisa Norris during radiotherapy treatment at the Beatson Oncology Centre, Glasgow in January 2006. Edinburgh: Scottish Executive. (ISBN 0-7559–6297–4)

Kemeny, J.G. et al. (1979). *Report of the President's Commission on the accident at Three Mile Island*. New York: Pergamon Press.

Kubota, R., Kiyokawa, K., Arazoe, M., Ito, H., Iijima, Y., Matsushima, H., and Shimokawa, H. (2001). Analysis of Organisation-Committed Human Error by Extended CREAM. *Cognition, Technology & Work*, 3, 67–81.

Lenorovitz, J.M. (1990). Indian A320 crash probe data show crew improperly configured aircraft. *Aviation Week and Space Technology, 132 (6/25/90)*, 84–85.

Lenorovitz, J.M. (1992). Confusion over flight mode may have role in A320 crash. *Aviation Week and Space Technology, 137 (2/3/92)*, 29–30.

Leveson, N.G. and Turner, C.S. (1992). *An investigation of Therac-25 accidents. UCI Technical Report* No. 92–108. University of California at Irvine.

Leveson, N.G. and Turner, C.S. (1993). An investigation of the Therac-25 Accidents. *Computer*, July, 18–41.

Linbeck, L. (2000). Special Commission on the 1999 Texas A&M Bonfire, Final Report, May 2, 2000.

Lions, J. (1996). Report of the Inquiry Board for Ariane 5 Flight 501 Failure, Joint Communication ESA-CNES, Paris, France.

Loeb, V. (2002). 'Friendly Fire' Deaths Traced to Dead Battery, Taliban Targeted, but U.S. Forces Killed. Washington Post, Sunday, March 24, 2002.

Monnier, A. (1992). Rapport preliminaire de la commission d'enquete administrative sur l`accident du Mont Sainte Odile du 20 Janvier 1992. Conference de Presse du 24 fevrier 1992, Ministere de L'equipement, du logement, des transports et de l'espace. Paris, France.

Moshansky, V.P. (1992). *Final report of the commission of inquiry into the Air Ontario crash at Dryden, Ontario*. Ottawa: Minister of Supply and Services, Canada.

Nelson, P.S. (2008). *STAMP Analysis of the Lexington Comair 5191 accident*. Lund University, Sweden: Unpublished MSc thesis.

National Transportation Safety Board (1984). *Eastern Air Lines Lockheed L-1011, N334EA Miami International Airport, FL, 5/5/83*. Report no. *AAR 84/04*, Springfield, VA: National Technical Information Service.

National Transportation Safety Board (1986a). *China Airlines B-747-SP, 300 NM northwest of San Francisco, CA, 2/19/85*. Report no. *AAR-86/03*, Springfield, VA: National Technical Information Service.

National Transportation Safety Board (1986b). *Delta Air Lines Lockheed L-1011–385–1, Dallas-Fort Worth Airport, TX, 8/2/85*. Report no. *AAR-86/05*, Springfield, VA: National Technical Information Service.

National Transportation Safety Board (1990). *Marine accident report: Grounding of the U.S. Tankship Exxon Valdez on Bligh Reef, Prince William Sound, near Valdez, Alaska,*

REFERENCES 253

March 24, 1989. Report no. NTSB/MAT-90/04. Springfield, VA: National Technical Information Service.

National Transportation Safety Board (1991). *Avianca, the airline of Colombia, Boeing 707–321B, HK 2016, Fuel Exhaustion, Cove Neck, NY, 1/25/90. Report no. AAR-91/04*, Springfield, VA: National Technical Information Service.

National Transportation Safety Board (1993). *US Air Flight 405 LaGuardia Airport, March 22, 1992. Report no. AAR-93/02*, Springfield, VA: National Technical Information Service.

National Transportation Safety Board (1996). Collision of Northeast Illinois Regional Commuter Railroad Corporation (METRA) Train and Transportation Joint Agreement School District 47/155 School Bus at Railroad/Highway Grade Crossing in Fox River Grove, Illinois, on October 25, 1995 (NTSB Rep. No. NTSB/HAR-96/02). Springfield, VA: National Technical Information Service.

National Transportation Safety Board (1997). *Grounding of the Panamanian passenger ship* Royal Majesty *on Rose and Crown shoal near Nantucket, Massachusetts, June 10, 1995*. (NTSB Rep. No. MAR-97/01). Springfield, VA: National Technical Information Service.

National Transportation Safety Board (2007). *Attempted takeoff from wrong runway Comair flight 5191, Bombardier CL-600–2B19, N431CA, Lexington, Kentucky, August 27, 2006* (NTSB/AAR-07/05). *(NTIS No. PB2007–910406)*. Springfield, VA: National Technical Information Service.

O'Connor, D.R. (2002). *Report of the Walkerton Inquiry*. Ontario Ministry of the Attorney General, Queen's Printer for Ontario.

Report Of The ANP/DPC Commission Of Investigation (2001). Analysis of the Accident With The Platform P-36. National Petroleum Agency and The Directorate Of Ports And Coast of the Brazilian Naval Command.

Rogers, W.P., et al. (1986). *Report of the Presidential Commission on the Space Shuttle Challenger Accident*. Washington, DC: Government Printing Agency.

Stephenson, A.G., et al. 2000. *Report on Project Management in NASA by the Mars Climate Orbiter Mishap Investigation Board*. NASA, March 13.

System Safety Message (2007). Potential For Erroneous Accuracy From Commercial And Military Global Positioning System (GPS) Receivers. C-E Lcmc Ground Precautionary Action (GPA) Message, GPA07–007. http://www.monmouth.army.mil/cecom/safety/system/safetymsg.htm

Toft, B. (2001). External Inquiry into the adverse incident that occurred at Queen's Medical Centre, Nottingham, 4th January 2001. Department of Health. PO Box 777, London SE1 6XH. www.doh.gov/qmcinquiry

U.S. Air Force Aircraft Accident Investigation Board Report (1999). RQ4A Global Hawk UAV, 98–2003, Edwards AFB, CA.

U.S. Department of Defense (1988). *Report of the formal investigation into the circumstances surrounding the downing of Iran Air Flight 655 on 3 July 1988*. Washington DC: Department of Defense.

U.S. House of Representatives Committee on Armed Services (1987). *Report on the staff investigation into the Iraqi attack on the USS Stark*. Springfield, VA: National Technical Information Service.

U.S. Nuclear Regulatory Commission (1985). *Loss of main and auxiliary feedwater at the Davis-Besse Plant on June 9, 1985. NUREG-1154.* Springfield, VA: National Technical Information Service.

BOOKS AND ARTICLES

Adams, M., Tenney, Y., and Pew, R.W. (1995). Situation awareness and the cognitive management of complex systems. *Human Factors, 37,* 85–104.

Amalberti, R. (2001). The paradoxes of almost totally safe transportation systems. *Safety Science, 37,* 109–126.

Ashby, W.R. (1956). *An introduction to cybernetics.* London: Chapman & Hall.

Baars, B.J. (ed.) (1992). *The experimental psychology of human error: Implications for the architecture of voluntary control.* NY: Plenum Press.

Bainbridge, L. (1987). Ironies of automation. In J. Rasmussen, K. Duncan, and J. Leplat (eds), *New technology and human error* (pp. 271–283). Chichester, UK: Wiley.

Barley, S. and Orr, J. (eds) (1997). *Between craft and science: Technical work in US settings.* Ithaca, NY: IRL Press.

Barnett, A. and Wang, A. (2000). Passenger mortality risk estimates provide perspectives about flight safety. *Flight Safety Digest, 19*(4), 1–12. Washington, DC: Flight Safety Foundation.

Baron, J. and Hershey, J. (1988). Outcome bias in decision evaluation. *Journal of Personality and Social Psychology, 54(4),* 569–579.

Becker, H.S. (1963). *Outsiders: Studies in the sociology of deviance.* New York: Free Press.

Billings, C. (1991). *Human-centered aircraft automation: A concept and guidelines. NASA Tech Memo 103885.* Springfield, VA: National technical Information Service.

Billings, C.E. (1996). *Aviation automation: The search for a human-centered approach.* Hillsdale, NJ: Lawrence Erlbaum Associates.

Boeing Product Safety Organization. (1993). *Statistical summary of commercial jet aircraft accidents: Worldwide operations, 1959–1992.* Seattle, WA: Boeing Commercial Airplanes.

Boring, E.G. (1950). *A history of experimental psychology (2nd Edition).* New York: Appleton-Century-Crofts.

Branlat, M., Anders, S., Woods, D.D., and Patterson, E.S. (2008). Detecting an Erroneous Plan: Does a System Allow for Effective Cross-Checking? In E. Hollnagel, C. Nemeth, and S.W.A. Dekker (eds), *Resilience engineering perspectives 1: Remaining sensitive to the possibility of failure.* Aldershot, UK: Ashgate, pp. 247–257.

Bransford, J., Sherwood, R., Vye, N., and Rieser, J., (1986). Teaching and problem solving: Research foundations. *American Psychologist, 41,* 1078–1089.

Brown, J.S., Moran, T.P., and Williams, M.D. (1982). *The semantics of procedures.* Xerox Palo Alto Research Center.

Brown, J.P. (2005a). Ethical dilemmas in health care. In M. Patankar, J.P. Brown, and M.D. Treadwell (eds), *Safety ethics: Cases from aviation, health care, and occupational and environmental health.* Burlington VT: Ashgate.

Brown, J.P. (2005b). Key themes in health care safety dilemmas. In M. Patankar, J.P. Brown, and M.D. Treadwell (eds), *Safety ethics: Cases from aviation, health care, and occupational and environmental health.* Burlington VT: Ashgate.

Bruner, J. (1986). *Actual minds, possible worlds*. Cambridge, MA: Harvard University Press.

Bryne, M.D. and Bovair, S. (1997). A working memory model of a common procedural error. *Cognitive Science, 21(1)*, 31–61.

Caplan, R., Posner, K., and Cheney, F. (1991). Effect of outcome on physician judgments of appropriateness of care. *Journal of the American Medical Association, 265*, 1957–1960.

Carroll, J.M., Kellogg, W.A., and Rosson, M.B. (1991). The Task-Artifact Cycle. In J.M. Carroll (ed.), *Designing interaction: Psychology at the human-computer interface*, Cambridge University Press, (p. 74–102).

Carroll, J.M. and Rosson, M.B. (1992). Getting around the task-artifact cycle: How to make claims and design by scenario. *ACM Transactions on Information Systems, 10(2)*, 181–212.

Cheng, P. and Novick, L. (1992). Covariation in natural causal induction. *Psychological Review, 99*, 365–382.

Chi, M.T.H., Glaser, R., and Farr, M. (1988). *The nature of expertise*. Lawrence Erlbaum Associates. Hillsdale: NJ.

Chopra, V., Bovill, J.G., Spierdijk, J., and Koornneef, F. (1992). Reported significant observations during anaesthesia: A prospective analysis over an 18-month period. *British Journal of Anaesthesia, 68*, 13–17.

Christie, N. (2004). *A suitable amount of crime*. London: Routledge.

Christoffersen, K. and Woods, D.D. (2002). How to make automated systems team players. In E. Salas (ed.), *Advances in human performance and cognitive engineering research*, Volume 2. St. Louis, MO, Elsevier Science, 1–12,

Christoffersen, K., Woods, D.D., and Blike, G.T. (2007). Discovering the Events Expert Practitioners Extract from Dynamic Data Streams: The mUMP Technique. *Cognition, Technology, and Work, 9*, 81–98

Clarke, L. and Perrow, C. (1996). Prosaic Organizational Failure. *American Behavioral Scientist, 39(8)*, 1040–1056.

Cook, R.I., McDonald, J.S., and Smalhout, R. (1989). *Human error in the operating room: Identifying cognitive lock up*. Cognitive Systems Engineering Laboratory Technical Report *89-TR-07*, Columbus, OH: Department of Industrial and Systems Engineering, The Ohio State University.

Cook, R.I. and O'Connor, M.F. (2004). Thinking about accidents and systems. In H. Manasse and K. Thompson (eds), *Improving medication safety*. American Society for Health-System Pharmacists, Bethesda, MD.

Cook, R.I. and Woods, D.D. (1994). Operating at the sharp end: The complexity of human error. In M.S. Bogner (ed.), *Human error in medicine*. Hillsdale, NJ: Lawrence Erlbaum Associates.

Cook, R.I. and Woods, D.D. (1996a). Implications of automation surprises in aviation for the future of total intravenous anesthesia (TIVA). *Journal of Clinical Anesthesia, 8*, 29s–37s.

Cook, R.I. and Woods, D.D. (1996b). Adapting to new technology in the operating room. *Human Factors, 38(4)*, 593–613.

Cook, R.I. and Woods, D.D. (2006). Distancing through Differencing: An Obstacle to Learning Following Accidents. In E. Hollnagel, D.D. Woods, and N. Leveson (eds), *Resilience engineering: Concepts and precepts*. Aldershot, UK: Ashgate, pp. 329–338.

Cook, R.I., Nemeth, C., and Dekker, S.W.A. (2008). What went wrong at the Beatson Oncology Centre? In E. Hollnagel, C. Nemeth, and S.W.A. Dekker (eds), *Resilience engineering perspectives 1: Remaining sensitive to the possibility of failure*. Aldershot, UK: Ashgate.

Cook, R.I., Potter, S.S., Woods, D.D., and McDonald, J.S. (1991-b). Evaluating the human engineering of microprocessor-controlled operating room devices. *Journal of Clinical Monitoring, 7,* 217–226.

Cook, R.I., Render M.L., and Woods, D.D. (2000). Gaps in the continuity of care and progress on patient safety. *British Medical Journal, 320,* 791–794, March 18, 2000.

Cook, R.I., Woods, D.D., and Howie, M.B. (1992). Unintentional delivery of vasoactive drugs with an electromechanical infusion device. *Journal of Cardiothoracic and Vascular Anesthesia, 6,* 1–7.

Cook, R.I., Woods, D.D., and McDonald, J.S. (1991-a). *Human performance in anesthesia: A corpus of cases*. Cognitive Systems Engineering Laboratory Technical Report *91-TR-03*, Columbus, OH: Department of Industrial and Systems Engineering, The Ohio State University.

Cook, R.I., Woods, D.D., and Miller, C. (1998). *A Tale of two stories: Contrasting views on patient safety*. National Patient Safety Foundation, Chicago IL, April 1998 (available at http://csel.eng.ohio-state.edu/blog/woods/archives/000030.html)

Cooper, J.B., Newbower R.S., and Kitz, R.J. (1984). An analysis of major errors and equipment failures in anesthesia management: Conditions for prevention and detection. *Anesthesiology, 60,* 43–42.

Cooper, J.B., Newbower, R.S., Long, C.D, and McPeek, B. (1978). Preventable anesthesia mishaps: A study of human factors. *Anesthesiology, 49,* 399–406.

Cordesman, A.H. and Wagner, A.R. (1996). *The lessons of modern war: Vol. 4: The Gulf War*. Boulder, CO: Westview Press.

Dauer, E.A. (2004). Ethical misfits: Mediation and medical malpractice litigation. In Sharpe, V.A. (ed), *Accountability: Patient safety and policy reform*, pp. 185–202. Washington DC: Georgetown University Press.

Dekker, S.W.A. (2002). *The field guide to understanding human error*. Aldershot, UK: Ashgate Publishing Co.

Dekker, S.W.A. (2005). *Ten questions about human error: A new view of human factors and system safety*. Mahwah, NJ: Lawrence Erlbaum Associates.

Dekker, S.W.A. (2006). *The field guide to understanding human error*. New edn, Aldershot, UK: Ashgate Publishing Co.

Dekker, S.W.A. (2007). *Just culture: Balancing safety and accountability*. Aldershot, UK: Ashgate Publishing Co.

De Keyser, V. and Woods, D.D. (1990). Fixation errors: Failures to revise situation assessment in dynamic and risky systems. In A.G. Colombo and A. Saiz de Bustamante (eds), *System reliability assessment*. The Netherlands: Kluwer Academic, 231–251.

Dorner, D. (1983). Heuristics and cognition in complex systems. In Groner, R., Groner, M., and Bischof, W.F. (eds), *Methods of heuristics*. Hillsdale NJ: Lawrence Erlbaum Associates.

Dougherty, E.M. (1990). Guest editorial. *Reliability Engineering and System Safety, 29,* 283–299.

Dougherty, E.M. and Fragola, J.R. (1988). *Human reliability analysis, a systems engineering approach with nuclear power plant applications.* New York: John Wiley.

Dunbar, K. (1995). How Scientists really reason: Scientific reasoning in real-world laboratories. In R.J. Sternberg and J. Davidson (eds), *Mechanisms of insight.* Cambridge, MA: MIT Press, pp. 365-395.

Edwards, W. (1984). How to make good decisions. *Acta Psycholgica, 56,* 5–27.

Endsley, M.R. (1995). Measurement of situation awareness in dynamic systems. *Human Factors, 37,* 65–84.

Es, G.W.H. van, Geest, P.J. van der, and Nieuwpoort, T.M.H. (2001). *Safety aspects of aircraft operations in crosswind (NLR-TP-2001–217).* Amsterdam, NL: National Aerospace Laboratory NLR.

Feltovich, P.J., Ford, K.M., and Hoffman, R.R. (1997). *Expertise in context.* MIT Press, Cambridge, MA.

Feltovich, P.J., Spiro, R.J., and Coulson, R. (1989). The nature of conceptual understanding in biomedicine: The deep structure of complex ideas and the development of misconceptions. In D. Evans and V. Patel (eds), *Cognitive science in medicine: Biomedical modeling.* Cambridge, MA: MIT Press.

Feltovich, P.J., Spiro, R.J., and Coulson, R. (1993). Learning, teaching and testing for complex conceptual understanding. In N. Fredericksen, R. Mislevy, and I. Bejar (eds), *Test theory for a new generation of tests.* Hillsdale NJ: Lawrence Erlbaum.

Feltovich, P.J., Spiro, R.J., and Coulson, R. (1997). Issues of expert flexibility in contexts characterized by complexity and change. In Feltovich, P.J., Ford, K.M., and Hoffman, R.R. (eds), *Expertise in context.* MIT Press, Cambridge, MA.

Ferguson, J. and Fakelmann, R. (2005). The culture factor. *Frontiers of Health Services Management, 22,* 33–40.

Fischer, U., Orasanu, J.M., and Montvalo, M. (1993). Efficient decision strategies on the flight deck. In *Proceedings of the Seventh International Symposium on Aviation Psychology.* Columbus, OH: April.

Fischhoff, B. (1975). Hindsight ≠ foresight: The effect of outcome knowledge on judgment under uncertainty. *Journal of Experimental Psychology: Human Perception and Performance, 1(3),* 288–299.

Fischhoff, B. (1977). Perceived informativeness of facts. *Journal of Experimental Psychology: Human Perception and Performance, 3,* 349–358.

Fischhoff, B. (1982). For those condemned to study the past: Heuristics and biases in hindsight. In D. Kahneman, P. Slovic, and A. Tversky (eds), *Judgment under uncertainty: Heuristics and biases.* Cambridge: Cambridge University Press.

Fischhoff, B. and Beyth, R. (1975). "I knew it would happen" Remembered probabilities of once-future things. *Organizational Behavior and Human Performance, 13,* 1–16.

Fischhoff, B. and Beyth-Marom, R. (1983). Hypothesis evaluation from a Bayesian perspective. *Psychological Review, 90(3),* 239–260.

Fitts, P.M. and Jones, R.E. (1947). *Analysis of factors contributing to 460 "pilot-error" experiences in operating aircraft controls. Memorandum Report TSEAA-694–12.* Dayton, OH: Aero Medical Laboratory, Air Materiel Command.

Flach, J., Hancock, P., Caird, J., and Vicente, K. (eds) (1996). *An ecological approach to human-machine systems I: A global perspective.* Hillsdale, NJ: Lawrence Erlbaum Associates.

Flach, J.M. and Dominguez, C.O. (1995). Use-centered design: Integrating the user, instrument, and goal. *Ergonomics in Design, 3,* 3, 19–24.

Flores, F., Graves, M., Hartfield, B., and Winograd, T. (1988). Computer systems and the design of organizational interaction. *ACM Transactions on Office Information Systems, 6,* 153–172.

Fraser, J.M., Smith, P.J., and Smith Jr., J.W. (1992). A catalog of errors. *International Journal of Man-Machine Studies, 37(3),* 265–393.

Gaba, D.M., Maxwell, M., and DeAnda, A. (1987). Anesthetic mishaps: Breaking the chain of accident evolution. *Anesthesiology, 66,* 670–676.

Gaba, D.M. and DeAnda, A. (1989). The response of anesthesia trainees to simulated critical incidents. *Anesthesia and Analgesia, 68,* 444–451.

Gaba, D.M. and Howard, S.K. (2002). Fatigue among clinicians and the safety of patients. *New England Journal of Medicine, 347(16),* 1249–1255.

GAIN (2004). *Roadmap to a just culture: Enhancing the safety environment.* Global Aviation Information Network (Group E: Flight Ops/ATC Ops Safety Information Sharing Working Group).

Garner S. and Mann, P. (2003). Interdisciplinarity: Perceptions of the value of computer supported collaborative work in design for the built environment. *International Journal of Automation in Construction, 12(5),* 495–499.

Gentner, D. and Stevens, A.L. (eds) (1983). *Mental models.* Hillsdale NJ: Lawrence Erlbaum Associates.

Gomes, J.O., Woods, D.D., Rodrigues de Carvalho, P.V., Huber, G., and Borges, M. (2009). Resilience and brittleness in the offshore helicopter transportation system: Identification of constraints and sacrifice decisions in pilots' work. *Reliability Engineering and System Safety, 94,* 311–319.

Gopher, D. (1991). The skill of attention control: Acquisition and execution of attention strategies. In *Attention and performance XIV.* Hillsdale, NJ: Lawrence Erlbaum Associates.

Gras, A., Moricot, C., Poirot-Delpech, S.L., and Scardigli, V. (1994). *Faced with automation: The pilot, the controller, and the engineer* (translated by J. Lundsten). Paris: Publications de la Sorbonne.

Guerlain, S., Smith, P.J., Obradovich, J.H., Rudmann, S., Strohm, P., Smith, J., and Svirbely, J. (1996). Dealing with brittleness in the design of expert systems for immunohematology. *Immunohematology, 12(3),* pp. 101–107.

Hart, H.L.A. and Honore, A.M. (1959). *Causation in the law.* Oxford: Clarendon Press.

Hasher, L., Attig, M.S., and Alba, J.W. (1981). I knew it all along: Or did I? *Journal of Verbal Learning and Verbal Behavior, 20,* 86–96.

Hawkins, S. and Hastie, R. (1990). Hindsight: Biased judgments of past events after the outcomes are known. *Psychological Bulletin, 107(3),* 311–327.

Heft, H. (2001). *Ecological psychology in context: James Gibson, Roger Barker and the legacy of William James's radical empiricism.* Mahwah, NJ: Lawrence Erlbaum Associates.

Herry, N. (1987). Errors in the execution of prescribed instructions. In J. Rasmussen, K. Duncan, and J. Leplat. (eds), *New technology and human error.* Chichester: John Wiley and Sons.

Hilton, D. (1990). Conversational processes and causal explanation. *Psychological Bulletin, 197(1),* 65–81.

Hirschhorn, L. (1993). Hierarchy vs. bureaucracy: The case of a nuclear reactor. In K.H. Roberts (ed.), *New challenges to understanding organizations*. New York: McMillan.

Hirschhorn, L. (1997). Quoted in Cook, R.I., Woods, D.D., and Miller, C. (1998). *A tale of two stories: Contrasting views on patient safety*. Chicago, IL: National Patient Safety Foundation.

Hoch, S.J. and Lowenstein, G.F. (1989). Outcome feedback: Hindsight and information. *Journal of Experimental Psychology: Learning, Memory and Cognition, 15(4)*.

Hochberg, J. (1986). Representation of motion and space in video and cinematic displays. In K.R. Boff, L. Kaufman, and J.P. Thomas (eds), *Handbook of human perception and performance: Vol. I*. New York: John Wiley and Sons.

Hoffman, R.R., Lee, J.D., Woods, D.D., Shadbolt, N., Miller, J., and Bradshaw, J. (2009). The Dynamics of Trust in Cyberdomains. *IEEE Intelligent Systems, 24(6)*, November/December, p. 5–11.

Hollnagel, E. (1991a). The phenotype of erroneous actions. In G.R. Weir and J.L. Alty (eds), *Human-computer interaction and complex systems*. London: Academic Press.

Hollnagel, E. (1991b). Cognitive ergonomics and the reliability of cognition. *Le Travail Humain, 54(4)*.

Hollnagel. E. (1993). *Human reliability analysis: Context and control*. London: Academic Press.

Hollnagel, E. (2004). *Barriers and accident prevention*. Aldershot, UK: Ashgate Publishing Co.

Hollnagel, E. (2009). The *ETTO principle: Efficiency-thoroughness tradeoff or why things that go right sometimes go wrong*. Aldershot, UK: Ashgate Publishing Co.

Hollnagel, E. and Woods, D.D. (2005). *Joint cognitive systems: Foundations of cognitive systems engineering*. Boca Raton FL: Taylor & Francis.

Hollnagel, E., Woods, D.D., and Leveson, N.G. (2006). *Resilience engineering: Concept and Precepts*. Aldershot, UK: Ashgate Publishing Co.

Hollister, W.M. (ed.) (1986). *Improved guidance and control automation at the man-machine interface*. AGARD Advisory Report No. 228, AGARD-AR-228.

Hughes, J., Randall, D., and Shapiro, D. (1991). CSCW: Discipline or paradigm? A sociological perspective. *European-Computer Supported Cooperative Work '91 (September)*, Amsterdam.

Hutchins, E. (1990). The technology of team navigation. In J. Galegher, R. Kraut, and C. Egido (eds), *Intellectual teamwork: Social and technical bases of cooperative work*. Hillsdale, NJ: Lawrence Erlbaum Associates.

Hutchins, E. (1995a). *Cognition in the wild*. Cambridge, MA: MIT Press.

Hutchins, E. (1995b). How a cockpit remembers its speeds. *Cognitive Science, 19*, 265–288.

Jagacinski, R.J., and Flach, J.M. (2002). *Control theory for humans: Quantitative approaches to modeling performance*. Mahwah, NJ: Lawrence Erlbaum Associates.

Johnson, P.E., Duran, A.S., Hassebrock, F., Moller, J., Prietula, M., Feltovich, P.J., and Swanson, D.B. (1981). Expertise and error in diagnostic reasoning. *Cognitive Science, 5*, 235–283.

Johnson, P.E., Moen, J.B., and Thompson, W.B. (1988). Garden path errors in diagnostic reasoning. In L. Bolec and M.J. Coombs (eds), *Expert system applications*. New York: Springer-Verlag.

Johnson, P.E., Jamal, K., and Berryman, R.G. (1991). Effects of framing on auditor decisions. *Organizational Behavior and Human Decision Processes, 50*, 75–105.

Johnson, P.E., Grazioli, S., Jamal, K., and Zualkernan, I. (1992). Success and failure in expert reasoning. *Organizational Behavior and Human Decision Processes, 53*, 173–203.

Karsh, B. (2004). Beyond usability for patient safety: designing effective technology implementation systems. *British Medical Journal: Quality and Safety in Healthcare, 13(5)*, 388–394.

Kelley, H.H. (1973). The process of causal attribution. *American Psychologist, 28*, 107–128.

Kelly-Bootle, S. (1995). *The computer contradictionary (2nd Edition)*. Cambridge, MA: MIT Press.

Klein, G.A. (1989). Do decision biases explain too much? *Human Factors Society Bulletin, 32(5)*, 1–3.

Klein, G.A. (1998). *Sources of power: How people make decisions*. Cambridge, MA: MIT Press.

Klein, G.A. and Crandall, B. (1995). The role of mental simulation in problem solving and decision making. In J. Flach, P. Hancock, J. Caird, and K. Vicente (eds), *An ecological approach to human-machine systems I: A global perspective*. Hillsdale NJ: Lawrence Erlbaum Associates.

Klein, G.A., Orasanu, J., and Calderwood, R. (eds) (1993). *Decision making in action: Models and methods*, Norwood NJ: Ablex.

Klein, G., Woods. D.D., Bradshaw, J., Hoffman, R.R., and Feltovich, P.J. (2004). Ten Challenges for Making Automation a "Team Player" in Joint Human-Agent Activity. IEEE Intelligent Systems, *19(6)*, 91–95.

Klein, G., Feltovich, P., Bradshaw, J.M., and Woods, D.D. (2005). Common Ground and Coordination in Joint Activity. In W. Rouse and K. Boff (ed.), *Organizational simulation*, Wiley, pp. 139–178.

Klein, G., Pliske, R., Crandall, B., and Woods, D. (2005). Problem Detection. *Cognition, Technology, and Work. 7(1)*, 14–28.

Kling, R. (1996). Beyond outlaws, hackers and pirates. *Computers and Society, 6*, 5–16

Krokos, K.J. and Baker, D.P. (2007). Preface to the special section on classifying and understanding human error. *Human Factors, 49(2)*, 175–176.

La Burthe, C. (1997). Human Factors perspective at Airbus Industrie. Presentation at International Conference on Aviation Safety and Security in the 21st Century. January 13–16, Washington, DC.

Lanir, Z. (1986). *Fundamental surprise*. Eugene, OR: Decision Research.

Lanir, Z., Fischhoff, B., and Johnson, S. (1988). Military risk taking: C_3I and the cognitive function of boldness in war. *Journal of Strategic Studies, 11(1)*, 96–114.

LaPorte, T.R. and Consolini, P.M. (1991). Working in Practice but not in Theory: Theoretical Challenges of High-Reliability Organizations. *Journal of Public Administration Research and Theory, 1*, 19–47.

Lauber, J.K. (1993). A safety culture perspective. Paper presented at Flight Safety Foundation's 38th Corporate Aviation Safety Seminar, April 14–16, Irving, TX.

Lerner, J.S. and Tetlock, P.E. (1999). Accounting for the effects of accountability. *Psychological Bulletin, 125*, 255–275.

Leveson, N. (2002). *System safety engineering: Back to the future*. Boston: MIT Aeronautics and Astronautics.

Lewis, C. and Norman, D.A. (1986). Designing for error. In D.A. Norman and S.W. Draper (eds), *User centered system design: New perspectives of human-computer interaction*, Hillsdale, New Jersey: Lawrence Erlbaum Associates, 411–432.

Lipshitz, R. (1989). "Either a medal or a corporal:" The effects of success and failure on the evaluation of decision making and decision makers. *Organizational Behavior and Human Decision Processes, 44*, 380–395.

Loftus, E. (1979). *Eyewitness testimony*. Cambridge: Harvard University Press.

Lubar, S.D. (1993). *History from things*. Washington DC: Smithsonian Inst Press.

Lützhöft, M. and Dekker, S.W.A. (2002). On your watch: Automation on the bridge. *Journal of Navigation, 55(1)*, 83–96.

Mach, E. (1905). *Knowledge and error*. Dordrecht: Reidel Publishing Company. (English Translation, 1976).

Marx, D. (2001). *Patient safety and the "just culture": A primer for health care executives*. Columbia University, New York.

Maurino D.E., Reason J.T., Johnston, N., and Lee, R.B. (1999). *Beyond aviation human factors*. Aldershot, UK: Ashgate Publishing Co.

McCarthy, J.C., Healey, P.G.T., Wright, P.C., and Harrison, M.D. (1997). Accountability of work activity in high-consequence work systems: Human error in context. *International Journal of Human-Computer Studies, 47*, 735–766.

McGuirl, J.M., Sarter, N.B., and Woods, D.D. (2009). See is Believing? The effects of real-time imaging on Decision-Making in a Simulated Incident Command Task. *International Journal of Information Systems for Crisis Response and Management, 1(1)*, 54–69.

McRuer, D. et al. (eds) (1992). *Aeronautical technologies for the twenty-first century*. Washington, DC: National Academy Press, 243–267.

Moll van Charante, E., Cook, R.I., Woods, D.D., Yue, Y., Howie, M.B. (1993). Human-computer interaction in context: Physician interaction with automated intravenous controllers in the heart room. In H.G. Stassen (ed.), *Analysis, design and evaluation of man-machine systems 1992*, Pergamon Press.

Monk, A. (1986). Mode errors: A user centred analysis and some preventative measures using keying-contingent sound. *International Journal of Man Machine Studies, 24*, 313–327.

Moray, N. (1984). Attention to dynamic visual displays in man-machine systems. In R. Parasuraman and D.R. Davies (eds), *Varieties of attention*. Academic Press.

Moray, N. and Huey, B. (eds) (1988). *Human factors research and nuclear safety*. Washington, DC: National Academy Press.

Morreim, E.H. (2004). Medical errors: Pinning the blame versus blaming the system. In Sharpe, V.A. (ed), *Accountability: Patient safety and policy reform*, pp 213–232. Washington DC: Georgetown University Press.

Murphy, R.R. and Woods, D.D. (2009). Beyond Asimov: The Three Laws of Responsible Robotics. *IEEE Intelligent Systems, 24(4)*, July/August, 14–20.

Murray, C. and Cox, C.B. (1989). *Apollo: The race to the moon*. New York: Simon and Schuster.

Neisser, U. (1976). *Cognition and reality: Principles and implications of cognitive psychology*. San Francisco, CA: W.H. Freeman and Company.

Newell, A. (1982). The knowledge level. *Artificial Intelligence, 18*, 87–127.

Norman, D.A. (1981). Categorization of action slips. *Psychological Review, 88*, 1–15.
Norman, D.A. (1983). Design rules based on analysis of human error. *Communications of the ACM, 26*, 254–258.
Norman, D.A. (1988). *The psychology of everyday things*. New York, NY: Basic Books.
Norman, D.A. (1990a). The 'problem' of automation: Inappropriate feedback and interaction, not 'overautomation.' *Philosophical Transactions of the Royal Society of London, B 327*, 585–593.
Norman, D.A. (1990b). Commentary: Human error and the design of computer systems. *Communications of the ACM, 33(1)*, 4–7.
Norman D.A. (1992). *Turn signals are the facial expressions of automobiles*. New York: Addison-Wesley.
Norman, D.A. (1993). *Things that make us smart*. Reading, Addison-Wesley: MA.
Obradovich, J.H. and Woods, D.D. (1996). Users as designers: How people cope with poor HCI design in computer-based medical devices. *Human Factors, 38(4)*, 574–592.
Orasanu, J.M. (1990). *Shared mental models and crew decision making*. CSL Report 46. Princeton: Cognitive Science Laboratory, Princeton University.
O'Regan, J.K. (1992). Solving the "real" mysteries of visual perception: The world as an outside memory. *Canadian Journal of Psychology/Revue Canadienne de Psychologie, 46(3)*, 461–488.
Patterson, E.S., Roth, E.M., and Woods, D.D. (2001). Predicting Vulnerabilities in Computer-Supported Inferential Analysis under Data Overload. *Cognition, Technology and Work, 3(4)*, 224–237.
Patterson, E.S., Roth, E.M., Woods, D.D., Chow, R., and Gomez, J.O. (2004). Handoff strategies in settings with high consequences for failure: Lessons for health care operations. *International Journal for Quality in Health Care, 16(2)*, 125–132.
Patterson, E.S., Cook, R.I., Woods, D.D., and Render, M.L. (2004). Examining the Complexity Behind a Medication Error: Generic Patterns in Communication. IEEE SMC Part A, *34(6)*, 749–756.
Patterson E.S., Woods, D.D., Cook, R.I., and Render, M.L. (2007). Collaborative cross-checking to enhance resilience. *Cognitive Technology and Work, 9(3)*, 155–162.
Patterson, E.S., Roth, E.M., and Woods D.D. (2010). Facets of Complexity in Situated Work. In E.S. Patterson and J. Miller (eds.), *Macrocognition metrics and scenarios: Design and evaluation for real-world teams*. Aldershot, UK: Ashgate.
Payne, J.W., Bettman, J.R., and Johnson, E.J. (1988). Adaptive strategy selection in decision making. *Journal of Experimental Psychology: Learning, Memory and Cognition, 14(3)*, 534–552.
Payne J.W., Johnson E.J., Bettman J.R., and Coupey, E. (1990). Understanding contingent choice: A computer simulation approach. *IEEE Transactions on Systems, Man, and Cybernetics, 20*, 296–309.
Perkins, D., Martin, F. (1986). Fragile knowledge and neglected strategies in novice programmers. In E. Soloway and S. Iyengar (eds), *Empirical studies of programmers*. Norwood, NJ: Ablex.
Perrow, C. (1984). *Normal accidents. Living with high-risk technologies*. New York: Basic Books.
Petroski, H. (2000). Vanities of the Bonfire. *American Scientist 88*, 486–490.

Pew, R.W., Miller, D.C., and Feehrer, C.E. (1981). *Evaluation of proposed control room improvements through analysis of critical operator decisions*. Palo Alto, CA: Electric Power Research Institute NP-1982.

Pidgeon, N. and O'Leary, M. (2000). Man-made disasters: Why technology and organizations (sometimes) fail. *Safety Science, 34*, 15–30.

Pohl, R.F. (1998). The effects of feedback source and plausibility of hindsight bias. *European Journal of Cognitive Psychology, 10(2)*, 191–212.

Rasmussen, J. (1985). Trends in human reliability analysis. *Ergonomics, 28(8)*, 1185–1196.

Rasmussen, J. (1986). *Information processing and human-machine interaction: An approach to cognitive engineering*. New York: North-Holland.

Rasmussen, J. (1990). The role of error in organizing behavior. *Ergonomics, 33(10/11)*, 1185–1199.

Rasmussen, J. (1997). Risk management in a dynamic society: A modeling problem. *Safety Science, 27(2/3)*, 183–213.

Rasmussen, J. and Batstone, R. (1989). Why do complex organizational systems fail? *Environment Working Paper No. 20*, The World Bank.

Rasmussen, J., Duncan, K., and Leplat, J. (eds) (1987). *New technology and human error*. Chichester: John Wiley and Sons.

Reason, J. (1990). *Human error*. Cambridge, England: Cambridge University Press.

Reason, J. (1993). The identification of latent organizational failures in complex systems. In J.A. Wise, V.D. Hopkin, and P. Stager (eds), *Verification and validation of complex systems: Human factors issues*. Springer-Verlag: Berlin. NATO ASI Series.

Reason, J. (1997). *Managing the risks of organizational accidents*. Aldershot, UK: Ashgate Publishing Co.

Reason, J. and Mycielska, K. (1982). *Absent minded? The psychology of mental lapses and everyday errors*. Englewood Cliffs, NJ: Prentice Hall.

Reason, J.T., Hollnagel, E., and Pariès, J. (2006). *Revisiting the "swiss cheese" model of accidents (EEC Note No. 13/06)*. Brussels: Eurocontrol.

Roberts, K.H. and Rousseau, D.M. (1989). Research in nearly failure-free, high-reliability organizations: Having the bubble. *IEEE Transactions in Engineering Management, 36*, 132–139.

Rochlin, G., LaPorte, T.R., and Roberts, K.H. (1987). The self-designing high reliability organization: Aircraft carrier flight operations at sea. *Naval War College Review, (Autumn)*, 76–90.

Rochlin, G. (1991). Iran Air Flight 655 and the USS Vincennes. In LaPorte, T. (ed.), *Social responses to large technical systems*. The Netherlands: Kluwer Academic.

Rochlin, G.I. (1993). Defining high-reliability organizations in practice: A taxonomic prolegomenon. In K.H. Roberts (ed.), *New challenges to understanding organizations*. (pp. 11–32). New York, NY: Macmillan.

Rochlin, G.I. (1999). Safe operation as a social construct. *Ergonomics, 42*, 1549–1560.

Roe, E. and Schulman, P. (2008). *High Reliability Management: Operating at the Edge*. Stanford, CA: Stanford University Press.

Rosness, R., Guttormsen, G., Steiro, T., Tinmannsvik, R.K., and Herrera, I.A. (2004). *Organisational accidents and resilient organizations: Five perspectives (Revision 1), Report no. STF38 A 04403*. Trondheim, Norway: SINTEF Industrial Management.

Roth, E.M., Bennett, K.B., and Woods, D.D. (1987). Human interaction with an "intelligent" machine. *International Journal of Man-Machine Studies, 27*, 479–525.

Roth, E.M. and Woods, D.D. (1988). Aiding human performance: I. Cognitive analysis. *Le Travail Humain, 51(1)*, 39–64.

Roth E.M., Woods, D.D., and Pople, H.E., Jr. (1992). Cognitive simulation as a tool for cognitive task analysis. *Ergonomics, 35*, 1163–1198.

Rouse, W.B. et al. (1984). *A method for analytical evaluation of computer-based decision aids (NUREG/CR-3655)*. Springfield, VA: National Technical Information Service.

Rouse, W.B. and Morris, N.M. (1986). On looking into the black box: Prospects and limits in the search for mental models. *Psychological Bulletin, 100*, 359–363.

Rudolph, J.W., Morrison, J.B., and Carroll, J.S. (2009). The Dynamics of Action-Oriented Problem Solving: Linking Interpretation and Choice. *Academy of Management Review, 34(4)*, 733–756.

Sandnes, D.L., Stephens, L.S., Posner, K.L., and Domino, K.B. (2008). Liability Associated with Medication Errors in Anesthesia: Closed Claims Analysis. *Anesthesiology, 109*, A770.

Sampson, J. (2000). The accepted airmanship standard is the stabilized approach. *Air Safety Week*, 13 November 2000.

Sarter, N.B. and Woods, D.D. (1991). Situation Awareness – A Critical But Ill-Defined Phenomenon. *International Journal of Aviation Psychology, 1(1)*, 45–57.

Sarter, N.B. and Woods, D.D. (1992). Pilot interaction with cockpit automation I: Operational experiences with the Flight Management System. *International Journal of Aviation Psychology, 2*, 303–321.

Sarter N.B. and Woods D.D. (1993). Pilot interaction with cockpit automation II: An experimental study of pilots' mental model and awareness of the Flight Management System (FMS). *International Journal of Aviation Psychology*, in press.

Sarter, N.B. and Woods, D.D. (1995). "How in the world did we get into that mode?" Mode error and awareness in supervisory control. *Human Factors, 37*, 5–19.

Sarter, N.B. and Woods, D.D. (1997). Teamplay with a Powerful and Independent Agent: A Corpus of Operational Experiences and Automation Surprises on the Airbus A-320. *Human Factors*, in press.

Sarter, N.B., Woods, D.D., and C. Billings, C. (1997). Automation Surprises. In G. Salvendy (ed.), *Handbook of human factors/ergonomics (2nd Edition)*. New York: Wiley.

Savage-Knepshield, P. and Martin, J.H. (2005). A Human Factors Field Evaluation of a Handheld GPS For Dismounted Soldiers. Proceedings of the Human Factors and Ergonomics Society 49th Annual Meeting. 26–28 September, Orlando, FL.

Schwenk, C. and Cosier, R. (1980). Effects of the expert, devil's advocate and dialectical inquiry methods on prediction performance. *Organizational Behavior and Human Decision Processes. 26*.

Schwid, H.A. and O'Donnell, D. (1992). Anesthesiologist's management of simulated critical incidents. *Anesthesiology, 76*, 495–501.

Seifert, C.M. and Hutchins, E. (1992). Error as opportunity: Learning in a cooperative task. *Human-Computer Interaction, 7(4)*, 409–435.

Sellen, A.J., Kurtenbach, G.P., and Buxton, W.A.S. (1992). The prevention of mode errors through sensory feedback. *Human-Computer Interaction, 7*, 141–164.

Senders, J. and Moray, N. (1991). *Human error: Cause, prediction, and reduction.* Hillsdale, NJ: Lawrence Erlbaum Associates.

Sharpe, V.A. (2003). Promoting patient safety: An ethical basis for policy deliberation. *Hastings Center Report Special Supplement 33(5)*, S1-S20.

Simon, H. (1957). *Models of man (Social and rational).* New York: John Wiley and Sons.

Simon, H. (1969). *The sciences of the artificial.* Cambridge, MA: MIT Press.

Singleton, W.T. (1973). Theoretical approaches to human error. *Ergonomics, 16,* 727–737.

Skegg, P.D.G. (1998). Criminal prosecutions of negligent health professionals: The New Zealand experience. *Med. Law Rev. 6,* 220–246.

Smith, E.E. and Goodman, L. (1984). Understanding written instructions: The role of an explanatory schema. *Cognition and Instruction, 1,* 359–396.

Snook, S.A. (2000). *Friendly fire: The accidental shootdown of US Black Hawks over Northern Iraq.* Princeton, NJ: Princeton University Press.

Spiro, R.J., Coulson, R.L., Feltovich, P.J., and Anderson, D.K. (1988). Cognitive flexibility theory: Advanced knowledge acquisition in ill-structured domains. *Proceedings of the Tenth Annual Conference of the Cognitive Science Society.* Hillsdale, NJ: Lawrence Erlbaum Associates.

Starbuck, W.H. and Milliken, F.J. (1988). Challenger: Fine-tuning the odds until something breaks. *Journal of Management Studies, 25,* 319–340.

Stech, F.J. (1979). *Political and Military Intention Estimation. Report N00014–78–0727.* Bethesda, MD: US Office of Naval Research, Mathtech Inc.

Stevens, S.S. (1946). Machines cannot fight alone. *American Scientist, 34,* 389–400.

Suchman, L. (1987). *Plans and situated actions. The problem of human machine communication.* Cambridge: Cambridge University Press.

Sutcliffe, K. and Vogus, T. (2003). Organizing for resilience. In K.S. Cameron, I.E. Dutton, and R.E. Quinn (eds), *Positive organizational scholarship*, 94–110. San Francisco: Berrett-Koehler.

Tasca, L. (1990). *The Social Construction of Human Error.* (Unpublished Doctoral Dissertation) Department of Sociology, State University of New York, Stony Brook.

Turner, B.A. (1978). *Man-made disasters.* London: Wykeham.

Tversky, A. and Kahneman, D. (1974). Judgment under uncertainty: Heuristics and biases. *Science, 185,* 1124–1131.

Vaughan, D. (1996). *The Challenger launch decision: Risky technology, culture and deviance at NASA.* Chicago: University of Chicago Press.

von Winterfeldt, D., and Edwards, E. (1986). *Decision analysis and behavioral research.* Cambridge: Cambridge University Press.

Wagenaar, W. and Keren, G. (1986). Does the expert know? The reliability of predictions and confidence ratings of experts. In E. Hollnagel, G. Mancini, and D.D. Woods (eds), *Intelligent decision making in process control environments.* Berlin: Springer-Verlag.

Wagenaar, W. and Groeneweg, J. (1987). Accidents at sea: Multiple causes and impossible consequences. *International Journal of Man-Machine Studies, 27,* 587–598.

Wasserman, D., Lempert, R.O., and Hastie, R. (1991). Hindsight and causality. *Personality and Social Psychology Bulletin, 17(1),* 30–35.

Watts-Perotti, J. and Woods, D.D. (2009). Cooperative Advocacy: A Strategy for Integrating Diverse Perspectives in Anomaly Response. *Computer Supported Cooperative Work: The Journal of Collaborative Computing, 18(2)*, 175–198.

Weick, K.E. (1990): The vulnerable system: An analysis of the Tenerife air disaster. *Journal of Management, 16(3)*, 571–593.

Weick, K.E. (1993). The collapse of sensemaking in organizations: The Mann Gulch disaster. *Administrative Science Quarterly, 38(4)*, 628–652.

Weick, K.E., Sutcliffe, K.M., and Obstfeld, D. (1999). Organizing for high reliability: Processes of collective mindfulness. *Research in Organizational Behavior, 21*, 13–81.

Westrum, R. (1993). Cultures with requisite imagination. In J.A. Wise, V.D. Hopkin, and P. Stager (eds), *Verification and validation of complex systems: Human factors issues*. Springer-Verlag: Berlin. NATO ASI Series.

Wiener, E.L. (1989). *Human factors of advanced technology ('glass cockpit') transport aircraft. Technical Report 117528*. Moffett Field, CA: NASA Ames Research Center.

Wiener, N. (1964). *God and Golem*. Cambridge, MA: MIT Press.

Wilkinson, S. (1994). The Oscar November incident. *Air Space*, 80–87 (February–March).

Winograd, T. and Flores, F. (1987). *Understanding computers and cognition*. Reading, MA: Addison-Wesley.

Winograd, T. and Woods, D.D. (1997). Challenges for Human-Centered Design. In J. Flanagan, T. Huang, P. Jones, and S. Kasif, (eds), *Human-centered systems: Information, interactivity, and intelligence*. National Science Foundation, Washington DC, July, 1997.

Wood, G. (1978). The knew-it-all-along effect. *Journal of Experimental Psychology: Human Perception and Performance, 4(2)*, 345–353.

Woods, D.D. (1982). Operating decision making behavior during the steam generator tube rupture at the Ginna nuclear power station. In W. Brown and R. Wyrick (eds), *Analysis of steam generator tube rupture events at Oconee and Ginna*. Institute of Nuclear Power Operations. 82–030. (Also, Westinghouse Research and Development Center Report 82–1C57-CONRM-R2.)

Woods, D.D. (1988). Coping with complexity: The psychology of human behavior in complex systems. In L.P. Goodstein, H.B. Andersen, and S.E. Olsen (eds), *Tasks, errors, and mental models*. New York: Taylor & Francis.

Woods, D.D. (1990a). Modeling and predicting human error. In J.L. Elkind, S. Card, J. Hochberg, and B. Huey (eds), *Human performance models for computer-aided engineering*. Boston, MA: Academic Press: Harcourt Brace Jovanovich.

Woods, D.D. (1990b). Risk and human performance: Measuring the potential for disaster. *Reliability Engineering and System Safety, 29*, 387–405.

Woods, D.D. (1991). The cognitive engineering of problem representations. In G.R. Weir and J.L. Alty (eds), *Human-computer interaction and complex systems*. London: Academic Press.

Woods, D.D. (1993). Research methods for the study of cognition outside of the experimental psychology laboratory. In G.A. Klein, J. Orasanu, and R. Calderwood (eds), *Decision making in action: Models and methods*. New Jersey: Ablex.

Woods, D.D. (1995a). Towards a theoretical base for representation design in the computer medium: Ecological perception and aiding human cognition. In J. Flach,

P. Hancock, J. Caird, and K. Vicente (eds), *An ecological approach to human-machine systems I: A global perspective*. Hillsdale NJ: Lawrence Erlbaum Associates.

Woods, D.D. (1995b). The alarm problem and directed attention in dynamic fault management. *Ergonomics, 38(11)*, 2371–2393.

Woods, D.D. (1996). Decomposing Automation: Apparent Simplicity, Real Complexity. In R. Parasuraman and M. Mouloula (eds), *Automation Technology and Human Performance*. Mahwah, NJ: Lawrence Erlbaum Associates.

Woods, D.D. (2003). *Creating foresight: How resilience engineering can transform NASA's approach to risky decision making*. US Senate Testimony for the Committee on Commerce, Science and Transportation, John McCain, chair. Washington, DC, 29 October 2003.

Woods, D.D. (2005). Creating Foresight: Lessons for Resilience from Columbia. In W.H. Starbuck and M. Farjoun (eds.), *Organization at the Limit: NASA and the Columbia Disaster*. pp. 289–308. Malden, MA: Blackwell.

Woods, D.D. (2006). Essential Characteristics of Resilience for Organizations. In E. Hollnagel, D.D. Woods, and N. Leveson (eds), *Resilience engineering: Concepts and precepts*. Aldershot, UK: Ashgate, pp. 21–34.

Woods, D.D. (2009). Escaping Failures of Foresight. *Safety Science, 47(4)*, 498–501.

Woods, D.D. and Branlat, M. (in press). How Adaptive Systems Fail. In E. Hollnagel, Woods, D.D., Paries, J., and Wreathall, J. (eds), *Resilience engineering in practice*. Aldershot, UK: Ashgate.

Woods, D.D. and Cook, R.I. (2006). Incidents: Are they markers of resilience or brittleness? In E. Hollnagel, D.D. Woods, and N. Leveson (eds), *Resilience engineering: Concepts and precepts*. Aldershot, UK: Ashgate, pp. 69–76.

Woods, D.D. and Dekker, S.W.A. (2000). Anticipating the Effects of Technological Change: A New Era of Dynamics for Human Factors. *Theoretical Issues in Ergonomic Science, 1(3)*, 272–282.

Woods, D.D. and Hollnagel, E. (2006). *Joint cognitive systems: patterns in cognitive systems engineering*. Boca Raton, FL: Taylor & Francis.

Woods, D.D. and Roth, E.M. (1988). Cognitive Systems Engineering. In M. Helander (ed.), *Handbook of human-computer interaction*. New York: Elsevier.

Woods, D.D. and Sarter, N. (2000). Learning from Automation Surprises and Going Sour Accidents. In N. Sarter and R. Amalberti (eds), *Cognitive engineering in the aviation domain*, Hillsdale NJ: Erlbaum, pp. 327–354.

Woods, D.D. and Sarter, N. (2010). Capturing the Dynamics of Attention Control From Individual to Distributed Systems. *Theoretical Issues in Ergonomics, 11(1)*, 7–28.

Woods, D.D. and Shattuck, L.G. (2000). Distant supervision – local action given the potential for surprise *Cognition, Technology and Work, 2*, 242–245.

Woods, D.D. and Watts, J.C. (1997). How Not To Have To Navigate Through Too Many Displays. In Helander, M.G., Landauer, T.K., and Prabhu, P. (eds), *Handbook of human-computer interaction (2nd Edition)*. Amsterdam, The Netherlands: Elsevier Science.

Woods, D.D., O'Brien, J., and Hanes, L.F. (1987). Human factors challenges in process control: The case of nuclear power plants. In G. Salvendy (ed.), *Handbook of human factors/ergonomics*. New York: Wiley.

Woods, D.D., Patterson, E.S., and Cook, R.I. (2006). Behind Human Error: Taming Complexity to Improve Patient Safety. In P. Carayon (ed.), *Handbook of human factors in health care*. Erlbaum (pp. 455–472).

Woods, D.D., Potter, S.S., Johannesen, L., and Holloway, M. (1991). Human interaction with intelligent systems. Volumes I and II. *Cognitive Systems Engineering Laboratory Technical Report 91-TR-01 and 02*, Columbus, OH: Department of Industrial and Systems Engineering, The Ohio State University.

Wright, D., Mackenzie, S.J., Buchan, I., Cairns, C.S., and Price, L.E. (1991). Critical incidents in the intensive therapy unit. *Lancet 338*, 676–678.

Xiao, Y., Hunter, W.A., Mackenzie, C.F., Jeffries, NJ, Horst, R., and LOTAS Group (1996). Task complexity in emergency medical care and its implications for team coordination. *Human Factors, 38(4)*, 636–645.

Yue, L., Woods, D.D., and Cook, R.I. (1992). Cognitive engineering of the human computer interface: Redesign of an infusion controller in cardiac anesthesiology. *Cognitive Systems Engineering Laboratory Technical Report 92-TR-03*, Columbus, OH: Department of Industrial and Systems Engineering, The Ohio State University.

Zelik, D., Patterson. E.S., and Woods, D.D. (2010). Measuring Attributes of Rigor in Information Analysis. In E.S. Patterson and J. Miller (eds.), *Macrocognition metrics and scenarios: Design and evaluation for real-world teams*. Aldershot, UK: Ashgate.

Zhang, J. and Norman, D.A. (1994). Representations in distributed cognitive tasks. *Cognitive Science, 18*, 87–122.

Zhang, J. (1997). The nature of external representations in problem solving. *Cognitive Science, 21(2)*, 179–217.

INDEX

accident analysis 212–14
accidents 43–6
accountability 225–6, 230–33
adaptation 191–6
   designers 195–6
   error 195
adaptation breakdowns 12
adaptive systems 38–9
Air Ontario Flight 1363 accident 52–4
anesthesia 24, 123–5, 231–2
Apollo 13 mission 162–4
artifacts 29, 155–8
Ashby's Law of Requisite Variety 11
astronomy 4–5
automation 29, 30, 144–54
   clumsy 146–53
   manual takeover 188–9
   over-automation 169–70

Bangalore aviation incident 182–3
barriers 41
blame-free systems 232–3
Boeing 757 landing incident 47–8
brittle tailoring 194–5
buggy knowledge 104–5
bureaucratic culture 49

cascading automated warnings 209–11
case studies
   Air Ontario Flight 1363 accident 52–4

anesthesia 123–4, 231–2
Boeing 757 landing incident 47–8
cascading automated warnings 209–10
chemical fire 219–20
coffee makers 62–3
Columbia space shuttle accident 42, 83–7, 129
DC-8 airliner coffee makers 62–3
Eastern Airlines L1011 56–8
Helios Airways B737 80-82
hypotension 113–14
Lexington Comair 5191 accident 70–72
maritime navigation 174–82
medication administration record (MAR) system 65–8
myocardial infarction 101–2
navigation at sea 174–82
risk homeostasis 76
Royal Majesty 174–82
take-off checklists 76–7
Texas A&M University bonfire tragedy 45–6
Webster, Noah 215–17
change
   and failure 246–7
   side effects of 12–13
chemical fire 219–24
clumsy automation 146–53, 191–6
coffee makers 62–3
cognition 151–3, 155–8

Cognitive Engineering 97–8
cognitive lockup 117–20
cognitive performance 110–11
cognitive systems 32
Columbia space shuttle accident 42, 83–92, 129
complexity 13, 248–9
computers 32–3
context sensitivity 166–7
contextual factors 26
control theory 69–77
coupling 62–5, 144
cross-scale interactions 93
culpability 227–30

data overload 143
DC-8 airliner coffee makers 62–3
decompensation incidents 24
design, of computer-based artifacts 29
distancing through differencing 89, 219–24
domino model 41–6

Eastern Airlines L1011 56–8
emergence 38
energy, risk as 36–8
erroneous actions 20–21
error analysis 212–14
error as information 215–24
error detection 26–9, 187–8
error recovery 26–8
error tolerance 27–8
errors 235–9, *see also* human error
escalation principle 151–2

failure 243–4, 246–7
failure sensitive strategies 10–11
feedback 187–8, 248–9
fixation 117–20
flight management computers (FMCs) 146–8
food preparation 47
fragmented problem-solving 90
fundamental surprise 47, 215–19

generative culture 49–50

goal conflicts 123–40
going sour scenarios 185–7

Helios Airways B737 80–82
heuristics 108–10
high reliability theory 77–82
hindsight bias 14–16, 33–4, 199–214, 242
human-computer interaction 152
human error 3–5, 19–20, 135–6, 239–41
human expertise, enhancement of 189–90, 247–8
Human Factors 30–31
hypotension 113–16, 136

iceberg model 44
Impact Flow Diagram 155–8
incidents 43–6
inert knowledge 107–8
interactive complexity 62–5

just culture 226–7, 233–5

keyhole property of computer displays 160–62
knowledge 101–4
  buggy knowledge 104–5
  cognitive performance 110–11
  inert 107–8
  oversimplifications 108–10
  technology change 105–6
knowledge calibration 106–7

latent failures 24–5, 50–59
learning opportunities 224
Lexington Comair 5191 accident 70–72
local rationality 16–17, 137–40

man-made disaster theory 46–50
maritime navigation 174–82
medication administration record (MAR) system 65–8
mental models 104–5
mindset 113–21
mode error 171–85
multiple standards 238
myocardial infarction 101–3

n-tuple bind 137–40
navigation at sea 174–82
neutral practitioner criteria 23, 211–12
normal accident theory 61–9
normative models 23, 207
nuclear power plants 23–4

operating rooms 148–51
operational complexity 187
organizational culture 49–50
outcome failures 21–2, 26–7
outcome knowledge 201–5
over-automation 169–70
oversimplifications 108–10

pathological culture 49
patterns of failure 246
performance 23
performance failures 21–2, 26–7
process assessment 205–11
process defects 22–3
Psychologist's Fallacy 244

Requisite Variety, Ashby's Law of 11
resilience 11, 93
Resilience Engineering 36, 38, 83–95
responsibility-authority double binds 130–35
risk, as energy 36–8
risk homeostasis 76
Royal Majesty 174–82

safety 12, 139, 245–6

second stories 241–2
sequence of events model 41–6
situation awareness 116–17
spatial organization 193
standards
    for performance assessment 237–8
    for process evaluation 206–11
Swiss cheese model 24–5, 50–59
system accidents 61, 69
system failures 28–9, 31–2, 35–9, 199–201
systemic vulnerabilities 244–5

take-off checklists 76–7
technical work, failure 243–4
technology change 105–6
technology, clumsy use of 143–53
technology change 143–5
test flight accident 184–5
Texas A&M University bonfire tragedy 45–6
Three Mile Island accident 215–19
tight coupling 62–5
TMI (Three Mile Island accident) 215–19
training 190

use-centered design 196

virtuality, penalties of 158–69

Webster, Noah 215–17
workload management 193